Developing Judgment

Developing Judgment

Assessing Children's Work in Mathematics

Jean Moon

HEINEMANN
Portsmouth, NH

HEINEMANN
A division of Reed Elsevier Inc.
361 Hanover Street
Portsmouth, NH 03801-3912

Offices and agents throughout the world

The author and publisher wish to thank those who have generously given permission to reprint borrowed material:

Figures I–1 and I–2 reprinted by permission of Jean Moon and Linda Schulman: *Finding the Connections: Linking Assessment, Instruction, and Curriculum in Elementary Mathematics* (Heinemann, A division of Reed Elsevier Inc., Portsmouth, NH, 1995).

Figure 1–1 from *A Question of Thinking: A First Look at Students' Performance on Open-ended Questions in Mathematics*, California Assessment Program. Copyright © 1989. Reprinted by permission of the California State Department of Education.

Library of Congress Cataloging-in-Publication Data

Moon, Jean.
Developing judgment : assessing children's work in mathematics / Jean Moon.
p. cm.
Includes bibliographical references.
ISBN 0-435-07140-8
1. Mathematics—Studying and teaching (Elementary)—Evaluation. 2. Mathematical ability—Testing. I. Title.
QA135.M61558 1997
372.7'.44—dc20 97-1577
CIP

Editors: Leigh Peake/Alan Huisman
Production: Vicki Kasabian
Text design: Joni Doherty
Cover design: Jenny Jensen Greenleaf

Printed in the United States of America on acid-free paper
01 00 99 98 97 DA 1 2 3 4 5

Contents

About This Book

In my work facilitating the introduction of alternative assessments into the mathematics curriculum, I have become increasingly aware that teachers all over the country are finding creative ways to form study groups. Math teachers/specialists in a school or a district frequently set up regular meetings to examine a topic of mutual interest. They identify a topic—alternative assessment, perhaps, or, more likely, one component of alternative assessment, such as student portfolios—and select appropriate resource material to guide their discussions. Their study therefore reflects the particulars of their teaching situations, and their teaching situations inform their study group work.

This book derives from a series of sessions in which a study group of elementary teachers and their principals discussed children's mathematical work in relation to alternative assessment. The study group included six teachers (grades one through five) from two elementary schools; the principals of the two schools; and a fourth-grade teacher from a neighboring district, who was coordinating a similar study group in her district while completing a year as a teacher fellow at the Center for Mathematics, Science, and Technology in Education at Lesley College. We met once a month throughout the 1995–96 school year.

During our sessions, we:

- Discussed children's responses to open-ended mathematical tasks.
- Identified and then applied criteria or performance indicators to the children's work.
- Built up our own capacity for expert judgment.

- Talked about what teachers look for in mathematical problems that they can use for assessment purposes.
- Discussed the role of informal classroom assessment in providing information that aids our understanding of teaching and learning.

The purpose of this book is fivefold:

1. To describe the next tier of work for teachers in alternative assessment and how that work can be applied to elementary mathematics.
2. To share the findings of a project study group engaged in this next tier of assessment.
3. To invite you to participate in the work of the project study group: to listen to their conversations and to think about the work of both the students and the study group members.
4. To give you the opportunity to build your expertise in judging student work by conducting each of the working sessions yourself, either individually or in conjunction with your colleagues.
5. To present a framework for organizing a study group around the work of your own students and those of your colleagues.

The book has been designed not only to be read, but to be used as a professional development resource. It is a guided journey through the process of developing judgment about student work.

In whatever way you use the book—alone, with one or two colleagues, or with your own study group—it invites action and reflection. Your active participation includes thinking about or discussing the student work provided here, listening to what the project study group members have to say about this same student work, and completing the suggested extension activities. Your reflective participation includes thinking about the ideas on teaching, learning, assessment, and mathematics expressed by the project study group members and similar ideas raised by you and your colleagues.

The book's "sessions" parallel the project study group's sessions. Each one deals with a major idea related to judging student work and how that idea is played out in instruction, curriculum, and assessment. Most of the sessions have six components:

1. Background information.
2. Session particulars, including goals and a recommended sequence.
3. Suggestions for conducting the session yourself.
4. Excerpts from the project study group's conversations.
5. My reflections on those conversations.
6. Ideas for additional opportunities to practice the assessment strategies presented.

Despite the shifting winds in politics and education, the commitment to using alternative assessment strategies in our children's classrooms continues to grow. Teachers and administrators are more consistently seeing the benefit of alternative assessment for diagnostic purposes and are becoming more respectful of this information in the teaching and learning process. More and more teachers and administrators are discovering that alternative assessment strategies, when embedded in everyday math lessons, have much to offer as a means by which to gain insights into teaching and document the progressive development of learning. And that's good news!

Acknowledgments

This book would not exist without the help of many generous people.

I am very grateful to the teachers and administrators in the project study group for their willingness to participate in an experiment, for their willingness to share their time and ideas and the work of their students, for their willingness to risk bringing new ideas into their classrooms and schools, and for their commitment to learning.

I am also grateful to Jo-Anne Rogers, my colleague in this adventure, who has my ongoing respect for her commitment to teachers, teaching, students, and learning.

Likewise, I want to acknowledge the Exxon Education Foundation, particularly Bob Witte and Ed Ahnert, for the ongoing opportunities I have had to listen to teachers in Exxon-supported K–3 math specialist projects throughout the United States talk about the challenges they face in integrating alternative assessment into their classrooms.

Leigh Peake, in her role in mathematics acquisitions at Heinemann, has been and continues to be a colleague of the best kind.

I am fortunate to have had Alan Huisman as my copy editor and as my sounding board for design and production concerns. Alan's expertise and thoughtfulness are a gift to any author.

I could never take on and complete any writing project without the support, participation, patience, and understanding of my family, Vicki and Doug. I gratefully, sincerely, and happily acknowledge the central role both have had in forming and shaping these thoughts and words. And I acknowledge, too, the role my extended family, particularly my best-friend cousins from Iowa, have had in shaping my appreciation for the art of conversation during summer evenings spent together on my grandmother's front porch.

INTRODUCTION

From Intuition to Expert Judgment

On page 220 of Susan Ohanian's *Garbage Pizza, Patchwork Quilts, and Math Magic* (1992), some teachers are talking about assessment:

> "There's a part of me that likes the 'cleanness' of percentages. Sometimes when I have to face the parents I wish I had the safety net provided by those percentages. Parents see my room and they are thrilled by all the projects. They say, 'Is this ever neat!' But sometimes there doesn't seem to be enough solid matter to grab on to. I mean, I can't tell them Johnny is a 93 percent. That has a nice definitive ring to it, doesn't it? Never mind Johnny; it makes me sound so precise, so scientific. But instead, I can show Johnny's work; I can describe a lot of his attitudes . . . but as we try to do this for twenty-five students I worry that we are falling into the old trap of jargon. I want to get away from developing a sort of jargon that just substitutes a verbal label for a numeric one." . . .
>
> "Right. But I worry about replacing the specific information of a 93 percent with a teacher's intuition. Most of what we do is, after all, intuition. But I'm afraid my intuition doesn't have a whole lot of validity with parents. Sometimes it doesn't have a whole lot of validity with me. . . ."
>
> "Yes, but all the traditional assessment assumptions of the old bell-shaped curve have been exploded. That really is no more 'objective' than what we're calling teacher intuition. Now we recognize—and demand—that assessment should give us information beyond a raw percentile score.

> This is hard for parents; . . . it's hard for us too. But what we're nervously calling intuition is a teacher who works six hours a day with the child. Maybe we need to stop being apologetic about our 'subjectivism' and start realizing that this is pretty well informed intuition."

Perhaps you have participated in a similar conversation or had a talk with yourself that touched on many of these points. The following questions are at the heart of many of our struggles related to alternative assessment:

- How do we develop a language about the work of children that can be viewed with the same validity as grades and percentages?
- How do we shift our emphasis from comparing children to being good diagnosticians of their work?
- How do we develop the skill to judge work that asks more of students than just the right answer?
- How can we develop confidence that what we are saying about students' development in math is valid, since this isn't something we learned in college?
- What evidence have we been using to support our judgments about student work and exactly how have we been using that evidence?

These questions are a natural progression in the work of alternative assessment. Once teachers understand assessment strategies such as pre- and postproblems, portfolios, and open-ended problems and are comfortable with the assessment terminology, actually judging students' work for purposes of assessment becomes the focus. An important part of this process of developing judgment is collaboration with your teacher colleagues. For teachers to develop good assessment abilities requires practice. The way to develop "expert judgment" in assessing student work is by "doing" assessment and then reflecting on what you've done.

The Next Steps

When I work with teachers, we spend a great deal of time poring over student work. We identify what mathematical knowledge the problem is asking the child to demonstrate. We talk about what learning attributes the problem is assessing and how we might revise it to make it an even better problem. We identify the range of mathematical abilities the problem might elicit from the students. We sort student work by whatever scoring rubric we have constructed together, and we discuss our bases for these judgments.

The richness of these sessions is the direct result of the availability of student work and the opportunity teachers have to talk about this work in depth. Everyone generally emerges with an understanding of what is required to move from intuition to analytic and expert judgment.

At the end of these sessions I am often asked questions like these:

- How can I replicate this experience for the teachers in my building or district, when I don't have access to student work like those you have provided?
- I need to have more practice in talking about student work; how can I do this?
- I am beginning to recognize some assumptions I have made about student work that I need to think about; how can I continue this process?

These are good questions. They suggest that teachers need to have greater access to "the real thing"—to focus discussions around a range of student work—in order to:

1. Practice interpreting not just their students' work, but the work of other students as well.
2. Develop an appropriate language for interpreting student work and then conveying the results to parents and to administrators, so that the work of both students and teachers is valued.
3. Get a sense of how their judgment of student work aligns with that of their colleagues.

Building expertise in examining student work is the next generation or tier of work teachers will be doing in assessment. And the way this will be done is through small study groups of teachers, subject specialists, and even building administrators who meet to look in depth at a topic of mutual interest.

This work will build on and at the same time create a greater understanding of what the teachers in Susan Ohanian's book call "intuition." Teachers, in conjunction with their teacher colleagues, will examine the evidence used to shape that intuition and in the process begin to take what has heretofore been a private hunch to the level of an informed judgment. For this kind of professional development, there is no set curriculum, no set menu of instructional activities, no predetermined checklist to be overlaid onto student work. Expertise can only be developed, constructed—much as we want our students to construct a facility with mathematics by doing mathematics, reflecting on their work, and practicing those areas that prove difficult until they "sense" their developing mathematical abilities alongside their ongoing development as students, individuals, and contributors to their class and school.

Developing a sense of the quality of one's educational performance is frequently compared to what happens to athletes and artists. Grant Wiggins (1993) comments,

> As all good athletes and coaches know, judgment and "anticipation" (perception of the unfolding situation) are essential elements of competence—so much so that players who are able to "read" the game can often compensate for skill deficiency. . . . What we must keep asking, then, is, What is the equivalent of the game in each subject matter? In other words, how is each subject "performed"? (p. 210)

My favorite performance analogy is the competitive diver. She begins every practice knowing which dives or diving techniques she needs to refine. Because she has taken part in numerous competitions, she does not depend solely on her coach to rate her performance. After executing a dive, she rarely has to be told whether the dive was good or poor. She just "knows." Over time she has gained an intuitive knowledge about her own performance that is invaluable to her as a learner and as a performer. The role of the coach is to put her in touch with the skills and abilities acquired through many, many practices.

No doubt these analogies are applied to alternative assessment so often because they convey just the right image. Developing a "sense" of one's growing expertise, be it in the gymnasium, the music room, the art studio, the garden, or the classroom, requires the ongoing cultivation of knowledge that goes beyond factual content. It is the "value added" dimension found in those who have mastered the facts and moved beyond to grasp what is often not seen at first glance—patterns of meaning acquired from early experiences, nuances, interconnections, good instincts, and an ability to analyze a situation.

Expert judgment, then, is an essential skill for teachers who are committed to using alternative assessment as a way to link their instruction and curriculum. This skill is difficult, if not impossible, to develop in isolation. The surest path to expert judgment is having repeated opportunities to do two things in tandem: (1) analyze student work according to performance indicators or criteria and (2) apply professional judgment during reflective conversations with colleagues about both the student work and the judgment of that work.

What Is Expert Judgment in the Service of Instruction?

The assessment cycle (see Figure I–1) is an iterative process—what is known about both the student's performance and the process itself is continually being refined by new information. And as the information on a student's performance is gathered, it also needs to be interpreted.

What does this student's response or series of responses tell me about his ability to organize and interpret data? At this point, the point of interpretation, the teachers' ability to make good judgments about student performance is critical not only to the iterative nature of the assessment cycle but to how well that cycle will help us make good instructional decisions.

Gathering information through a variety of tools or strategies—open-ended problems, pre- and postassessments, portfolios, interviews—will not automatically help us understand the progressive development of a student's ability. Gathering the information is a necessary first step, it's true, but that information needs to be interpreted according to common criteria in order to make it meaningful or useful. Interpreting, or "judging," student products is therefore the most critical step in the assessment process; it puts student work in perspective and documents a progressive pattern of development.

Collected information, or "evidence," needs to be interpreted in a couple of different ways. First, how did a student do relative to a specific task? (*What has Emma demonstrated in this particular project about her understanding of symmetry?*) Answering this question is important because doing so provides a glimpse of a student's ability at a particular moment in time. Second, how is a student's understanding of a mathematical idea developing over time? (*How is Emma*

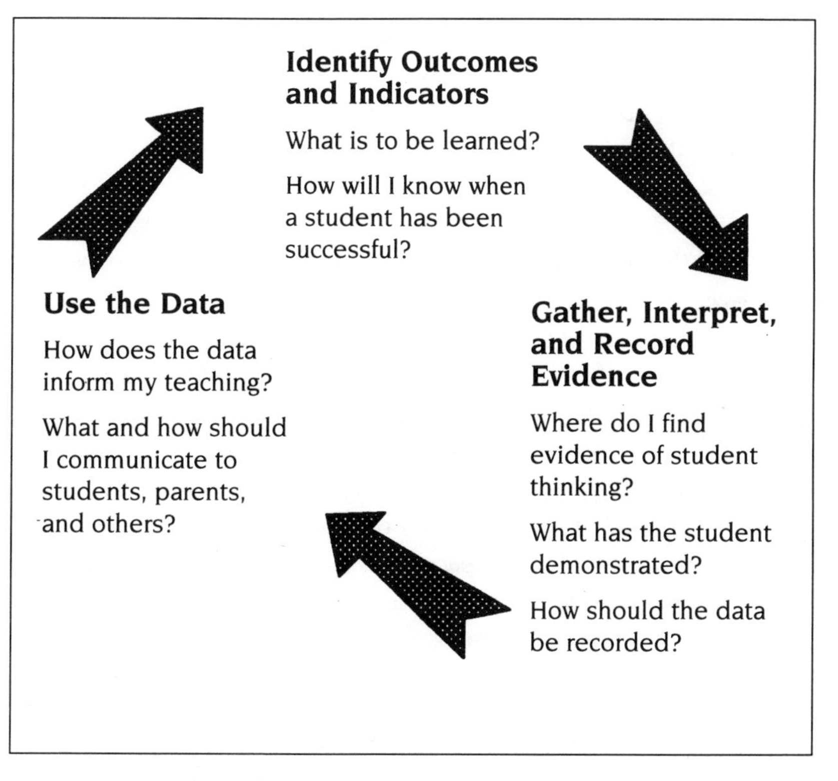

FIGURE I–1. Model of the assessment cycle.

doing in her understanding of symmetry?) The cumulative response to this question, assembled after repeated opportunities to judge, to evaluate, a student's work, provides a composite picture of her development. The ability to answer each question is essential to understanding the progressive development of student thinking.

Wiggins points out that judgment does not involve the unthinking application of rules or rubrics. He reminds us of John Dewey's dictum that "knack, tact, cleverness, insight, and discernment" are all part of the act of judging. Judging well does not come willy-nilly. Expertise in judgment comes from repeated practice in applying a set of standards or criteria to a specific situation. When standards and criteria intersect with student work, evaluation emerges. A decision needs to be made: how closely did a student meet the standard or the criterion in question?

Teachers, parents, and administrators have been drawn to alternative assessment because of a committed belief that what is important is not just coming up with a correct answer but "understanding" what is being taught. As Wiggins again reminds us, correct answers, unexamined, can mask misunderstanding, while incorrect answers, when the thinking behind them is explained, can reveal deep insight. The application of experienced, expert judgment on the part of teachers in concert with content standards, performance standards, and performance indicators ensures that student work is not just collected but examined to determine what the student understands.

A teacher once told me she had come to realize that children not only need to be able to make independent decisions about how to attack a math problem, they also need time to discuss their work, to share their thinking and their strategies with one another, to ask themselves questions like *What did I see that Roger didn't? What didn't I see that Sarah did? Why did Paul think this information was important and I didn't?* As a result of creating time in her math classes for discussion, this teacher has noticed her students becoming more comfortable communicating about their work and the thinking that went into it. She has also noticed that they seem willing to take more risks in trying different approaches to solving problems.

I believe her observation conveys a truth as well about how a sense of understanding is built up among teachers about the process of judging student work: it comes through repeated opportunities to discuss that work with other teachers. Again, performance analogies are helpful in stressing this point. Student work is in many ways like the work of artists, athletes, and cooks: much of its success depends on an intuitive understanding of the craft involved, an understanding that evolves through constant practice, collaboration, and reflection. In other words, students, artists, athletes, cooks get a "feel" for their work—colors, smells, perspective, placement, balance of ingredients—

by sharing their work with others—coaches, teachers, members of the audience, colleagues. This "feel" for quality can be challenging to describe but is clearly put into action nonetheless. (There is an excellent discussion of professional ways of knowing in Donald Schon's "Professional Knowledge and Reflective Practice" [1989].)

Good assessment practices depend not only on a well-developed intuitive understanding of a process (*how* one evaluates student work), but also on objective, public criteria by which a value is assigned to student work. The criteria move assessment from subjective interpretation to reasoned, professional judgment supported by evidence. The two go hand in hand and address the concern one of the teachers expresses in the passage from Susan Ohanian's book: "But I worry about replacing the specific information of a 93 percent with a teacher's intuition." *Developing Judgment* explores the process of turning intuition into professional judgment.

Why Are Open-Ended Problems Important?

Increasingly, teachers today are incorporating *open-ended problems* into their mathematics classrooms as a strategy to help their students apply math content different from standard algorithms. Open-ended problems also provide opportunities for students to explain how they solved a problem. When I ask teachers what comes to mind when they encounter the term *open-ended problems,* I generally get three responses:

- They have more than one answer, whereas traditional math problems have one right answer.
- They can be solved in a number of ways; students can arrive at a solution in reasonable ways other than using a standard algorithm.
- They give students a chance to communicate their thinking about how they solved the problem.

What is an open-ended mathematics problem?

How is an open-ended problem in mathematics different from the kinds of questions traditionally found in textbooks or on standardized tests?

In order to include open-ended problems in your own classroom, what changes do you need to make in your pedagogy?

Additionally, open-ended mathematics problems:

- Encourage students to think mathematically, to practice problem-solving strategies, and to develop their conceptual understanding of mathematical topics.
- Allow students to do mathematics in a context that is meaningful to them, such as collecting data on birthdays or pets or families.
- Are accessible to all students because they allow for a range of mathematical abilities.

Perhaps it is the opportunity students have to explain their thinking that has focused so much attention on open-ended problems as important connections linking instruction, curriculum, and assessment. Teachers tell wonderful stories about how listening to or reading student descriptions of their thinking provides a real "aha" about student learning and about their own teaching. It is becoming increasingly clear to teachers that instruction does not end with assessment. Good assessment and good instruction work together. Thought-provoking problems provide opportunities for children to present evidence of their mathematical thinking and decision making.

In *Finding the Connections* (1995), Linda Schulman and I give some examples of problems originally formed in the more traditional closed style and then re-formed into open-ended questions (see Figure I–2). The term *closed* is an appropriate contrast to the attributes just listed for open-ended problems. When a student is directed to focus on one right answer, further investigation or greater exploration is artificially closed to the student. The message conveyed is that she has completed the task by finding the right answer. The thinking process ends as she

What open-ended problems have you found that are particularly good ones for your students? What makes them good?

Closed	**Open**
Show a graph about favorite books and ask, "How many people chose biographies as their favorite books?"	Show a graph about favorite books and ask, "What does this graph tell you?"
Show a picture of a rectangle and ask, "What is the area of this figure?"	Ask, "Why would a gardener want to know the area of her garden? What other measurements might she want to know?"
Ask, "What is the value of the 3 in the number 3,472?"	Ask, "Imagine that you had a job numbering the pages in books. Your first book starts on page 1 and ends on page 462. How many times will you write the digit 3?"
Say, "Mr. Chang buys 8 plants and one pot. Each plant cost $5 and the pot cost $12. What was the total cost?"	Say, "Margo solved a problem. She multiplied two numbers and then added 12 to the product. Write a story problem that Margo could answer using this arithmetic."
Show two towers, one that is 8 cubes high and one that is 4 cubes high. Ask, "How many blocks should you move from this tower to this tower so the towers will be the same height?"	Say, "There are two towers of different heights. They are both built with the same size blocks. Write directions so that someone will know how to decide how many blocks to move so that the towers will be the same height."

FIGURE I–2. Examples of closed and open-ended problems.

Does the quote below sound familiar? Do you find yourself thinking and saying similar things?

"I do a lot more of letting the children follow through on someone's inappropriate response, having everybody discuss why the response isn't reasonable. A few years ago I'd say, No, that isn't right, *and show them the right thing to do. I wasn't letting everyone see why it didn't work, why it didn't make sense. This is all very different from when I went to school. Then it was very much arithmetic and very exact."*

moves on to the next problem. If that next problem is a repetition of the previous one in terms of the abilities needed to solve it, any need to communicate her thinking decreases even further; the emphasis becomes rote procedures, not building conceptual understanding.

Of course, just because a mathematics problem has all the attributes we've talked about, it still may not be a good problem for your students. Jan Mokros, Susan Jo Russell, and Karen Economopoulos (1995) suggest that a good mathematical problem is one that works in action. Often, developing a problem is a trial-and-error process: testing, reflecting, revising, again and again. When curriculum developers and teachers look for good problems, they ask questions like those in Figure I–3. Many of the ideas behind these questions reflect the attributes previously identified as important to open-ended mathematics problems, but there are some additional ideas there as well.

Traditional algorithm-dependent math problems associate mathematics with good memorization skills rather than with good thinking or reasoning skills. A traditional math class can be a dreaded experience for students whose learning styles do not include fluency in memorization.

As a student I was told I did not have to understand math; all I had to do was learn and apply the rules. For some students, myself included, learning the rules is not enough—the rules have to have

- Does solving this problem lead to work with significant mathematical ideas and relationships?
- Does the activity lead students to consider important mathematical ideas that reach beyond the particular result?
- Are there different ways into this problem, so that students with different strengths, needs, and experiences will be able to engage with some aspect of the problem?
- Is the problem interesting to a wide range of students?
- As students become involved in the problem, are they grabbed by the mathematics in the problem, or do nonmathematical aspects of the investigation tend to take center stage?
- Do the constraints of the problem provide enough direction and structure without overly restricting the ways in which students might think about the problem?
- Does the problem lead to satisfying closure for the students? Do they feel they have arrived somewhere when they are finished?
- Does the problem tend to support writing about, talking about, constructing, and representing student ideas?
- Are the necessary materials available and manageable?

FIGURE I–3. Questions to ask when evaluating whether a problem is good or not. From *Beyond Arithmetic: Changing Mathematics in the Elementary Classroom* by J. Mokros, S. Russell, and K. Economopoulos. Copyright © 1995 by Dale Seymour Publications. Reprinted by permission.

meaning. When the *why* of a rule is this important, rote memorization becomes a difficult task.

Memorization in mathematics is not the ultimate goal; fluency is. Clearly, it is important that students be able to provide the answer to 8 × 5. But do they need to be able to do it instantly? Is speed necessary to demonstrate an understanding of mathematics? Is there a relationship between computational speed and mathematical competence?

The notion behind number fluency is that a child should be able to take numbers apart and put them back together in a variety of ways, with a great deal of concentrated thinking. Achieving a deeper understanding of numbers and their relational parts is a much more powerful ability than one's speed in manipulating them. For example, in the 8 × 5 example, a student may figure out that if he adds 8 five times he will get 40. Adding 8 five times may not be efficient, but it conveys a basic understanding of number relationships that memorizing number facts does not.

Evidence

The introduction of open-ended problems is a critical first step in integrating assessment strategies into the classroom and making the connection between assessment and instruction. Open-ended tasks provide more robust opportunities for students to apply mathematical ideas. As a result, teachers have many more occasions to gather evidence of their students' thinking and to make instructional decisions on the basis of that evidence.

In this context, *evidence* is data that contribute to an understanding of students' development of mathematical concepts and provide public documentation of that development. There are many sources of evidence of student progress. They include students' written work samples, notes from interviews, a videotape of students engaged in a cooperative activity, recorded descriptions or notes taken by teachers observing students working on an open-ended problem, and students' written or spoken self-assessment.

After evidence is gathered, it needs to be evaluated in some way: *What do my observations tell me about Javier's sense of numbers? Is his sense of numbers in agreement with the way this important mathematical ability has been defined in his school or his class?* These questions are answered by laying performance standards and performance indicators on top of notes, descriptions, and student work samples in order to judge how a particular student is progressing.

To be more than opinion, evidence has to be measured against a common public standard. Identifying performance standards, performance indicators, and scoring rubrics is therefore an essential step if evidence is to be meaningful to teachers, students, and parents. Figure

I–4 is an example of a primary-grade number sense performance standard—the ability to understand what numbers mean—and performance indicators showing how well the standard is being met. (For a more complete discussion of the role of performance standards and performance indicators, see Moon and Schulman 1995.)

What other performance standards are part of an elementary student's developing number sense? How would you describe a progressive range of development for that performance standard? What would constitute evidence for you as a teacher that a student has achieved a well-developed sense of numbers?

Evidence is a key concept in understanding how assessment is connected to classroom instruction. Evidence connects teachers with what is central to teaching—evaluation. Teachers need to have a clear understanding of what their students know; only then can they make good instructional decisions. Unless teachers have evidence of learning that goes beyond test scores, unless they draw inferences from the evidence they have collected, the task of describing in very concrete language—language that makes sense to students and parents—what students *know* and *can do* in mathematics is very difficult. It may, by default, come back to, *How is this child doing in comparison with the other students in the class?*

Knowing and Doing Mathematics

Knowing and *doing* mathematics are integral concepts in the National Council of Teachers of Mathematics (NCTM) curriculum and evaluation standards but not necessarily easily definable ones. Elizabeth Badger (1992) describes what the standards intend by the phrase "knowing and doing mathematics" this way:

> When [the standards] define mathematical knowledge by listing different kinds of knowing, they are not suggesting a laundry list of discrete abilities but different facets of the same ability. For example, one doesn't understand concepts and then solve problems, one understands concepts by

Content Standard: Number Sense

One performance standard revealing a knowledge of what is meant by number sense: *The ability to understand what numbers mean*

Performance indicators I can observe in student behavior:

- *beginning ability*: recognizes that numbers represent a quantity
- *developing ability*: recognizes when a number is larger or smaller than another number
- *competent performance*: transfers knowledge of the magnitude of numbers to beginning number operations in addition

FIGURE I–4. A performance standard and related performance indicators.

> solving problems. One doesn't communicate mathematical ideas as well as reason about mathematics; rather, through communication one refines one's mathematical reasoning. This sense of wholeness reflects the general goal of mathematics education—what the curriculum standards refer to as "mathematical power." (p. 41)

When a teacher has multiple sources of information from a variety of open-ended problems, all of which contribute to understanding a student's ability in a particular mathematical concept, that teacher can then make a valid judgment or evaluation about the student's development. To some degree, student evaluation requires the kind of expert judgment demonstrated by such notable fictional detectives as Miss Marple and Sherlock Holmes. Good detectives gather evidence from many sources. Then, on the basis of the evidence they have collected, they make valid inferences that eliminate certain suspects, introduce others, and slowly suggest a reasonable scenario about how the crime was committed and who committed it.

So it is in teaching: ongoing efforts to collect evidence of student growth support a teacher's ability to make inferences on the basis of that evidence. From these inferences, teachers, students, parents, and administrators are able to determine what a student knows about doing mathematics. Teachers often have "hunches" about students who do not do well on end-of-chapter tests but show in other performance contexts that they really know mathematics. This is neither unusual nor unexpected. Clearly, one kind of evaluation method may not tell the whole story about a student; more evidence is needed, different sources of evidence are required.

Patterns in Student Performance

An important point to remember is that a perfect assessment, like a perfect standardized test, does not exist. One assessment problem or set of problems cannot tap the complex abilities we associate with mathematics. A more realistic approach is to look for patterns in student performance over time using an array of different kinds of problems in different contexts—students working in small groups, pairs of students working together, students working alone with problems some of which unfold over time and some of which can be completed in one class period. An important function of curriculum and instruction is to provide many learning experiences. It then becomes assessment's job to guide the ways in which patterns of information provided by those experiences are organized and interpreted in order to be meaningful to students, teachers, parents, and administrators.

In the process of examining different kinds of student work, particularly with other colleagues, a number of important things are going to happen for the teachers who participate in these discussions and for the students of these teachers:

1. Teachers will become more aware of the complexity of mathematical ideas and the relationships among those ideas; their own understanding of mathematics will grow.
2. Teachers will become more aware of their own expectations and biases about how students learn mathematics.
3. Teachers will become better diagnosticians of student work because they will have an increased clarity about, and will build a shared vocabulary with which to discuss, mathematical relationships and the variety of ways those relationships can be expressed.
4. Teachers will become clearer about their instructional goals—what it is they want to accomplish with their students.
5. Teachers will become clearer about what it is they value in teaching and learning mathematics. Depending on the logistics established for discussing student work, teachers from one grade level may hear teachers from other grade levels identify what they value as well. In this way, the conceptual relationships across grade levels will also become clearer.
6. Teachers will be able to share with their students this process for defining and then revisiting what it is they value in teaching and learning mathematics. Students can be asked to address, individually and in groups, such questions as *What is expected in math class? What does a good performance look like? How will I know when I am successful? Why was I more successful today than when I did this problem last week?*

Valuing Conversations with Colleagues

It is difficult to turn professional intuition into expert judgment in isolation. It is common practice for doctors and nurses to consult one another about a patient's treatment. Lawyers and judges talk with one another about interpretations of the law and judicial precedents. The Supreme Court justices work together to produce majority and dissenting opinions on the cases they review.

But the teaching profession has been slow to adopt a collaborative model of decision making and consultation. Schools do not typically set aside time for teachers to meet by grade level or across grade level to discuss professional practice. Although there are ample examples of teachers seeking one another's advice informally, five minutes here, ten minutes there, few school administrations provide time for professional collaboration. Ironically, strategies for shared problem solving and decision making among students in the classroom are becoming more common. Why not a similar model for teachers?

Those developing and promulgating alternative assessment make a point of emphasizing the importance of teacher collaboration as a critical tool in connecting assessment with curriculum. The richness of children's mathematical work on open-ended tasks and investigations is a wonderful spur to conversation. These conversations can lead to greater knowledge about students' progressive development of mathematical understanding and the role of instruction and curriculum in shaping that understanding.

SESSION 1

Creating Conversations Around Student Work

Background Information

Developing the judgment necessary to evaluate student work requires practice in conjunction with your colleagues. It is through these guided conversations that you will build a common language for diagnosing student work, build an analytic framework for judging student work, reflect on your teaching, and determine mathematical lessons that will provide the best opportunities for students to demonstrate district and state outcomes.

Often when teachers work together in small groups to rate student work they use assessment rubrics they have had no part in constructing. By contrast, the learning strategy for our project study group was both to construct a language for discussing student work and to practice setting criteria for judging that work before worrying about scoring it. Each meeting included lengthy, purposeful conversations about samples of children's mathematical activities (some of which were from the participants' own classrooms); based on these conversations, we practiced constructing criteria, finding evidence of those criteria, and then using that evidence to judge student work. We did not begin with a predetermined set of criteria or a generic scoring rubric.

From your reading, your inservice training, and/or your study of alternative assessment, you may recall that there are a number of approaches to assessing student work using some type of rubric. (The word *rubric* comes from the Latin *rubrica,* which means *red earth.* The Romans marked something of importance with red earth, *rubric terra.*) Rubrics are used primarily to help teachers and students mark or record student progress. Some provide a numeric score, while others yield a narrative profile of student progress based on descriptive categories. Some models do both. Rubrics generally fall into these categories:

- Generic holistic scoring rubrics (see Figure 1–1). In holistic scoring, a numeric rating is assigned to a student's work based on viewing that work as a "whole." To some, the benefit of this kind of rubric is that it does not evaluate a task per se, but considers a broad range of behavior associated with successfully accomplishing that task. A generic holistic scoring rubric can be constructed for individual

Level	Standard to be achieved for performance at specified level
6	**Fully achieves the purposes of the task, while insightfully interpreting, extending beyond the task, or raising provocative questions.** Demonstrates an in-depth understanding of concepts and content. Communicates effectively and clearly to various audiences, using dynamic and diverse means.
5	**Accomplishes the purposes of the task.** Shows clear understanding of concepts. Communicates effectively.
4	**Substantially completes the purposes of the task.** Displays understanding of major concepts, even though some less important ideas may be missing. Communicates successfully.
3	**Purposes of the task not fully achieved; needs elaboration; some strategies may be ineffectual or not appropriate; assumptions about the purposes may be flawed.** Gaps in conceptual understanding are evident. Limits communication to some important ideas; result may be incomplete or not clearly defined.
2	**Important purposes of the task not achieved; work may need redirection; approach to task may lead away from completion.** Presents fragmented understanding of concepts; results may be incomplete or arguments may be weak. Attempts communication.
1	**Purposes of the task not accomplished.** Shows little evidence of appropriate reasoning. Does not successfully communicate relevant ideas; presents extraneous information.

FIGURE 1–1. Example of a generic holistic scoring rubric, from the California Assessment Program (1989).

schools or school districts or can be imported from other districts and research organizations.

- Analytic scoring rubrics (see Figure 1–2). In this scoring approach, points are awarded for achieving specific components of an overall task (e.g., effectively communicating a solution or collecting data for the problem in appropriate ways).
- Descriptive rubrics (see Figure 1–3). Descriptive rubrics are constructed around progressive "descriptions" or benchmarks of student performance related to standards. (These descriptions generally are not translated into numbers.) For example, if a standard being emphasized at a grade level or across several grade levels is problem solving in mathematics, a descriptive rubric "describes" the development of abilities (performance indicators) associated with particular facets of problem solving. Suppose a facet of problem solving receiving instructional emphasis is a student's ability to formulate questions about the problem to be solved; the descriptive rubric breaks down that specific indicator—formulating questions—into a developmental progression of descriptions. These descriptions should fit along a continuum of beginning, developing, and achieving behavior. Other terminology is often substituted for *beginning, developing,* and *achieving*. The central idea is to state descriptively a range of developing

Applies appropriate mathematical solution to a problem:	0: No evidence of a solution strategy. 1: Evidence of a partial solution to the problem or solution offered is inappropriate. 2: Solution is both appropriate and complete.
Describes steps taken to solve problem:	0: No or little description of solution steps. 1: Description is incomplete, but shows partial understanding. 2: Description is complete.
Uses appropriate mathematical language in solution description:	0: Language is unclear or inappropriate. 1: Language contains appropriate terms, but usage is inconsistent. 2: Language consistently contains appropriate mathematical terms.
Conjectures about other appropriate solution strategies:	0: No evidence of other solution. 1: Conjectures made, but they are incomplete or inappropriate. 2: Alternative solution(s) is/are complete and appropriate.

FIGURE 1–2. Example of an analytic scoring rubric.

Activity: ______________________ Observation Dates:

Student: ______________________ ______________________

Collects, Organizes, and Displays Data

Collecting Data

Makes little attempt ______________________

Devises a plan, does not complete it ______________________

____

Devises and completes plan ______________________

____

Other observations ______________________

Classifying Information

Sorts information ______________________

____

Sorts information and builds categories ______________________

____

Decides what categories are useful to the task ______________________

____

Other observations ______________________

Representing Information

Makes efforts to translate information (from narrative to symbol or picture) ______________________

____

Translates and labels the representations in meaningful ways ______________________

____

Constructs representations to display key data in several ways ______________________

____

Other observations ______________________

FIGURE 1–3. Example of a descriptive rubric.

student performance behavior associated with an identified standard.

- Task-specific scoring rubrics (see Figure 1–4). Task-specific rubrics may be constructed either before students complete the task or afterward, and they may result in either numeric scoring or narrative documentation. The purpose is to focus on the problem at hand and the knowledge and skills demonstrated by students in providing an appropriate solution to that particular problem.

Ideally, the rubrics you use will be determined by the kind of information you need at a specific time about your students' developing mathematical abilities. If you want to get a sense of how students compare with one another, it may make more sense to depend on a

Date: ____________________

Student: ____________________

Estimation Problem:

Evidence of Progress:

Use of a Strategy:

Planning:

Executing:

Revising:

Estimation Skills:

Appropriate degree of accuracy:

Checks for reasonableness:

Number Sense:

Correct computation:

Understanding of ratio/proportion:

Representation of Data:

Notes for Instruction:

FIGURE 1–4. Example of a task-specific scoring rubric.

holistic or an analytic rubric. On the other hand, if you are more interested in looking closely at the process of your students' thinking and their overall development on specified mathematical abilities, the better choice is a descriptive rubric.

Because the work of this study group was twofold—to develop expertise in judging student work and to apply that information to instructional decisions—the best tools for us were descriptive rubrics that were task-specific.

Session Particulars

Goal

To analyze a mathematical task, Water Needed for Camping (see Figure 1–5), and to use this information to identify performance indicators that can serve as criteria for assessing students' written responses to it.

Sequence of activities

1. Analyze the assessment task. What is the Water Needed for Camping problem asking students to do? What knowledge and skills would a student need to be able to demonstrate in order to be successful in solving the problem?
2. Sort the student work samples for this problem provided at the end of this book. Sort student work into clusters that represent similar student performances in relation to identified performance criteria.
3. Reflect on the analysis and sorting process.

Suggestions

Before reading through the project study group's conversations as they discussed the Water Needed for Camping problem, become more familiar with the problem yourself. Reflect on both the mathematical and process skills required of students in order to solve it successfully. Answer the same question I posed to the project study group: *What knowledge and skills would a student need to be able to demonstrate in order to be successful in solving this problem?*

If you are using this book in connection with a study group, separate into working groups of two or three people and discuss this question for at least ten or fifteen minutes. Sometimes conversations like this are concluded prematurely. Try to push your conversations beyond the obvious points by asking additional questions: *How would your response to this question be different if the students were fifth graders instead of fourth graders? Which of the NCTM standards does this knowledge or action reflect?*

A group of 8 people are all going camping for 3 days and need to carry their own water. They read in a guide book that 12.5 liters are needed for a party of 5 people for 1 day. Based on the guide book, what is the minimum amount of water the 8 people should carry all together?

Explain your answer.

FIGURE 1–5. The Water Needed for Camping problem. Activity by Dr. Judith A. Arter. From A *Toolkit for Professional Developers: Alternative Assessment*. Copyright © 1994. Reprinted by permission of the publisher, Northwest Regional Educational Laboratory.

Project Study Group Discussion

The project study group members separated into three groups of three members each to discuss the Water Needed for Camping work samples. The groups worked for twenty minutes or so, analyzing the task on the basis of the question What knowledge and skills would a student need to be able to demonstrate in order to be successful in solving this problem? *The groups then sorted the twelve student samples into piles according to the level of demonstrated performance. The members made these judgments on the basis of whatever mathematical knowledge or skills they identified in their discussions. When they finished, we talked about what they had done.*

JM: Let's begin by looking at the problem. What does this question ask of a student? What mathematical knowledge must a student have in order to solve the problem successfully? What skills and abilities would a student need to have? *[After a few moments of silence]* Why don't I start. For example, I think a student has to be able to select the appropriate number operations in this problem. In other words, the students are going to make some kind of decision about if and when they should add, multiply, subtract, or divide.

Lucy: They also need to know that a liter is a unit used for measuring liquid, because that might throw them totally: "What is a liter?"

Sharon: Yes, "I can't do this, because I don't get that word." Maybe they would not be able to make that connection. They need to be able to read enough to comprehend the problem.

Bonnie: They need to know how to sort and organize information in order to set up the problem appropriately.

Maxine: And there has to be some facility with communication, being able to talk about what you do. As I looked through these samples, a number of the children represented what they did through numbers, but they didn't really write about it and describe it.

Kate: Right, part of the solution is communicating your thinking, having an ability to describe what you did when you solved the problem. Clearly, some of these students think that showing your arithmetic is explaining your answer. Are these samples from different schools?

JM: Yes. They were done by a cross-section of students who may or may not be at the same school with the same teacher.

Francine: I would like to go back to Kate's point about showing your work and explaining your answer. Many children think, I used these numbers to solve this problem, what more is there to say?

Vicki: Right, I can understand that response and I certainly have heard it. Maybe the "explain" part of the problem has been poorly worded. I think "explain your thinking" or "tell how you got your answer" puts kids on a different path than "explain your answer." *[There is a general murmur of agreement]*

Jo-Anne: Kids are going to have to know if their answer is reasonable or not reasonable in relation to the information given in the problem and the question the problem is asking. For instance, if this problem were asking about the amount of water a group of twenty-four people would need then that's three times the number of people, so a reasonable answer should be in proportion to that.

JM: Why doesn't each group describe their sorting process.

Maxine (Group 1): We had three piles. One pile, we felt, had a clear understanding. They decided on the correct operations and their computations were correct. The second pile demonstrated a beginning understanding but didn't follow through, didn't take it far enough. The third pile didn't have an understanding of what the numbers represented, or perhaps they misread the problem, but they weren't able to use the numbers to figure out how to solve the problem.

JM: Okay. Now if you were to articulate the criteria that you used to guide your sorting process, what would you say?

Maxine: The question we used was, Could they understand the relationship of the numbers in order to use them to solve the problem?

JM: The relationship among the numbers in the problem?

Maxine: Right.

JM: How about another group?

Gail (Group 2): Well, we basically described what we expected of responses in each category—low, middle, and high. Our problem was that in the top category, as we defined top, we had one minus. So it wasn't quite the top, but a little bit down. We looked at how the students used language to explain the steps in solving the problem, how well they showed how they understood their number operations and demonstrated number sense, and we included the reasonableness of the answer.

JM: So you really had three criteria for sorting the student work: evidence of problem solving, evidence of understanding number operations, and reasonableness of the answer?

Bonnie (Group 2): Yes. We felt that responses in the second group showed a sense of operations and used sequential problem solving but maybe missed a step or concept. And for those in the third group, there was little evidence of knowledge of operations, little written language, and no sequential steps.

Jo-Anne (Group 3): We put only two responses in our top category, one that met the criteria fully and one that partially did. The things we were looking for were whether they were able to articulate their reasoning and whether they were able to compute accurately—were they able to follow through in their reasoning and come to an accurate solution and answer? For example, the responses in the middle category showed a basic understanding of the problem, an understanding to a certain degree, but they didn't follow through. The students got halfway through the steps and got lost. But they had some understanding of where they were and were able to articulate it somewhat, through drawings or words. And the students in our third category went through a lot of motions, but they really didn't know what they were doing. They didn't seem to know what to do with eight people as opposed to five.

JM: For your group, Bonnie, was it confusing in your sorting process to be using three criteria simultaneously?

Bonnie: Maybe it was a little, because sometimes a student would be successful in demonstrating one of the criteria but not the other two. Accuracy with computation did not necessarily mean a student went about setting up the problem correctly. Sample 10 [*p. 113*] is an example of what I mean. The computation is accurate, but the student hasn't grasped the right relationships.

JM: I think it can be difficult to gather evidence on a number of criteria at the same time. This is a point we need to revisit. Does anyone have any other observations about this process of sorting the student work on the basis of criteria you and your partners developed?

Kate: Before I could begin to think about sorting anything, I had to solve the problem myself. My partners did the same thing, but

I didn't want to see how they solved it. I was really curious about how I would approach it, to get my own hold on somebody else's thinking. And it was interesting that we each did it differently. And trying to explain to one another how we did it, that was interesting and gave us some kind of grounding, an entry point for looking at the students' work. At least it did for me. It was hard.

Gail: I had to do that too. I mean, I was reading them and I said, I've got to figure out what this is. Then I could look at the students' process.

Vicki: Let me ask a question. What about the child who didn't show his work, the numbers, but wrote about it *[Sample 3, p. 99]*?

Kate: That's the one I loved the best, because I like language. She said, "I multiplied," and so forth.

Jo-Anne: But you know what? She articulated the last half of it thoroughly, but the first half she didn't.

Vicki: Are you penalizing styles?

JM: Are we looking at how she communicated the steps involved in solving the problem?

Gail: It actually says, explain your answer.

JM: But I think a lot of what is intended by that phrase depends on the culture of the classroom. In some classrooms, just showing your work, the computation used to solve the problem, would be enough of an explanation.

Sharon: We do a math challenge almost every morning. They want to know right away if they are right. They have the problem solved, but they have to show me the methods that they used. Some of them draw. They have a variety of different methods that they use to arrive at the answers.

JM: Your examples give us some insight into the culture you are creating in your classroom around explanation. That is, you seem to be setting expectations that your students will share their work publicly and that their mathematical explanations will involve a description of the method—steps—used to solve the problem, not just the answer by itself or just the computation used to produce an answer. So in a problem like the water problem, which asks the students to explain their answer, you would have established expectations that *explain* means more than show your work.

Bonnie: Something I noticed when the groups were explaining their sorting process was how all of us were pretty much in agreement when a similar criterion was being used, even though we didn't know a child's classroom situation. Our ratings were fairly close, which I thought was very interesting.

JM: You're right, during conversations like these, the judgments are usually not that far apart. Again, it seems to me that this is a result of the tacit knowledge we've developed but have not always shared publicly.

Lucy: I have a question on Sample 11 [*p. 115*]. I couldn't follow the thinking. But the student got it right. I just thought, maybe I've lost something here.

Sharon: I think we see this in the classroom all the time. This child started off—I mean there is no written language here so I don't know—but he started off with some kind of figures and then got to a point where obviously 12.5 got divided by 5. And then the 2.5 got multiplied by 4, which we couldn't figure out. Why did he multiply it by 4? We agreed that sometimes when children are working together in an unstructured way, quote, helping each other, unquote, everyone is working and really intent and moving along in the right direction, when suddenly the person next to them gets an answer. So the child says, Oh, you got sixty? and then looks at his own work and thinks, How can I get sixty, my partner got sixty?

Vicki: Clearly, in this example, we don't have enough information to determine where the child got his answer. It would be interesting to find out why. You know, go back to that child and ask him, Why did you multiply times four? Because it might have been leading in another direction and then something made him suddenly jump to adding fifty to the ten to equal sixty. I think this happens because the children are just working next to each other, they're not really collaborating. Perhaps in this process children get pushed to a place they are not ready to be. We think it's good to have them working together because it's supportive and all that, but sometimes it might really not be helpful to them.

Lucy: Maybe we are getting a little too scattered in what we do in classrooms, maybe there is a logical way of ordering things.

Vicki: I don't think it is a matter of being too scattered; children need the opportunities to test out their thinking. For me the challenge is to try and figure out how to keep track of all the different ways they are going. Just look at what these students did with this problem, they all took a little different approach to it. It's amazing!

JM: All of these comments in some way seem to relate to the phrase *explain your answer.* For those students who do go beyond showing their work, *explain your answer* means describe what you were thinking when you solved this problem and then what you actually did to solve it. These sorts of directives—explain, tell all you know, and so forth—are supposed to provide us with a window into student thinking so that we don't focus solely on the answer.

Francine: I understand all that. But when I first looked at this work, my attention went directly to the answers, who got the right answer. Once I had done that, then I could let myself look at how an answer was obtained.

Kate: And I couldn't look at the answers until I had worked the problem, seen things more as a whole. That's interesting.

Reflections

Group 1 described their sorting process this way: "One pile, we felt, had a clear understanding. They decided on the correct operations and their computations were correct. The second pile demonstrated a beginning understanding but didn't follow through, didn't take it far enough. The third pile didn't have an understanding of what the numbers represented, or perhaps they misread the problem, but they weren't able to use the numbers to figure out how to solve the problem." The criterion this group used to guide them was, "Could [the students] understand the relationship of the numbers in order to use them to solve the problem?"

This description of their process highlights an important principle for making judgments when assessing student work: *Are what you are stating as criteria for evaluating student work in fact what you are using to evaluate that work?* In other words, are you judging what you say you are judging? If not, it's classic apples and oranges: the stated principle is not the principle put into action. This point cannot be stressed enough.

In this session of our project study group, there were such inconsistencies. The stated criterion for Group 1 was students' knowledge of the mathematical relationships suggested in the Water Needed for Camping problem and their ability to select appropriate number operations on the basis of that knowledge. However, in the process of applying that criterion, Group 1 began to stretch or expand their stated criterion to include accuracy. For example, when Maxine described the most well-developed student work, she said those students had a clear understanding because they decided on the correct operations *and* their computations were correct. When she described the least-developed work, she indicated these students did not demonstrate an understanding of what the numbers represented or they misread the problem because they weren't able to use the numbers to figure out how to solve the problem; she didn't mention computational accuracy.

In thinking through the way in which the teachers in this particular work group applied their assessment criterion, another question comes to mind: can we assume that accurate computations, or even the selection of certain number operations, is evidence that a student understands the mathematical relationship in a problem? I'm not sure we can always make that assumption, even though the hoary charge to students to "show your work" encourages us to make it.

In the Water Needed for Camping problem, the telling evidence for how well a student understands the mathematical thinking comes when students are asked to explain their answers. In order to solve this problem successfully, a student needs to grasp that since 12.5 liters was the total for 5 people for 1 day, he needs to find out the relationship between 12.5 liters for 5 people for 1 day of camping and x number of liters for 8 people for 3 days. Accurate computations may or may not demonstrate a grasp of these mathematical relationships or proportional thinking. A student's stated relationship may be on target, but her computation may still be inaccurate. The reverse may happen as well: the computation may be successful, but a student's conceptual understanding may be shaky.

When sorting student work to determine the range of responses and identify performance indicators or criteria on which to judge that work, you should spend time looking at the student work before rushing to sort it or identify criteria. A couple of moments spent with the work while asking yourself, *What do I notice about this work? What stands out for me?* helps you see the patterns that will be helpful in the judging process and in finding the evidence to support your judgments. Criteria for judging can be set either before work is sorted or afterward. That is, you can sort it first using very general criteria and then sort it again when the criteria you want to use become clearer and more precise. You will know you have reached a point of clarity when you are able to put the criteria into words without a struggle.

The portion of the study group's conversation I've included suggests how frequently instructional issues arise in the course of discussing student work. The following excerpt touches on instruction as much as it does assessment:

> We agreed that sometimes when children are working together in an unstructured way, quote, helping each other, unquote, everyone is working and really intent and moving along in the right direction when suddenly the person next to them gets an answer. So the child says, Oh, you got sixty? and then looks at his work and thinks, How can I get sixty, my partner got sixty?

The other group members could easily identify with this dilemma. While the direction of the conversation shifted away from the task at hand at this point, it was a natural and worthwhile shift. It provided an opportunity to make ongoing connections between instruction and assessment. Similar shifts in conversation took place throughout our year together and helped us clarify our thinking related to classroom management and communication with parents.

More Ideas for You or Your Study Group

1. Earlier you were asked to identify what knowledge and skills a student would have to understand and apply in order to be successful in solving the Water Needed for Camping problem. Using one student ability you identified, such as the ability to communicate one's mathematical thinking, sort the student work samples for this problem according to how well knowledge of that ability is demonstrated. Establish guidelines for the sorting process. For example, you may not want to set up any expectations beforehand; let each working group determine their own guidelines about the number of piles and how sorting criteria will be determined. If you are working in a study group format, separate into working groups of two or three. After you all have finished sorting and discussing, ask each group to report on what happened and how they made decisions about what criteria to use to guide them.
2. Identify evidence from each student sample that supports your judgment about which pile that sample belongs in. Continually check the "fit" of the evidence with the student ability guiding the process, as well as the content of the evidence. Can the evidence used to guide your judgment be found in the sample of student work, or is your judgment based on an inference or "guess"? Evidence should be present in the student work itself, not inferred. (For example, if you say that a student's computation is inconsistent, give examples of inconsistencies obtained from that student's written work.)
3. Duplicate the sequence of activities identified for this session, but substitute work from your students or from the students of a colleague. (Make sure the problem asks them to share their thinking about how they went about solving it.)

SESSION 2

The Sorting Process

Background Information

At first glance, sorting student work seems to involve simple decisions based on stated criteria conforming to a single rubric. This simplicity, however, can be deceiving. As discussed in the Creating Conversations session, the challenge is to be consistent, to use the criteria you say you are going to use and not others. Also, it is important to be clear about what the criteria mean. Criteria that are unclear or too general will make your sorting inconsistent. This consistency between saying and doing is part of building your expertise in judging student work. The sorting process requires conscientious attention to a few principles:

- Evaluate work on the basis of the agreed-on criteria. Check in with others from time to time to see whether everyone has the same understanding of the criteria being used.
- Do not compare student responses with one another. The task is to compare each piece of student work with agreed-on criteria or with a more formal rubric.
- Evaluate what is presented in the product you are evaluating, not what you think the student meant to write, compute, or draw.
- Keep a record of the misconceptions students appear to have that are causing them to make errors in their work. Is there a pattern to these misconceptions that requires some reteaching and relearning?

The convention is to begin by placing student work into three piles. One pile is for excellent work—it meets or perhaps exceeds whatever criteria have been specified. Another pile is for work somewhere between excellent and underdeveloped—it has met some of the criteria but not all. The third pile generally contains

The sorting "to dos" can be combined into what I call the CAMP principle:

- Be Consistent; don't make Comparisons.
- Don't make Assumptions.
- Identify Misconceptions.
- Identify Patterns.

work that meets few or none of the criteria and suggests that the originator needs to relearn and practice the mathematical concepts involved.

When there are a great many papers in the middle pile, that pile is often re-sorted into two, for a total of four piles, and the descriptions of the piles altered accordingly. For example:

Pile 1: The work meets few or none of the criteria.

Pile 2: The originator of the work needs substantial practice.

Pile 3: The originator of the work needs selected practice.

Pile 4: The work meets or exceeds the criteria.

Looking at patterns in student work within each category as well as across the categories yields very good diagnostic information about your students' development in relation to the key mathematical ideas and skills contained within the assessment task. This kind of sort also provides you with good information on which to base instructional decisions regarding those students who are not demonstrating evidence of reasonable progress.

Generic holistic scoring rubrics may have five or six performance levels that include or exclude student work on the basis of the completeness of the response. The number of levels in a formal rubric is arbitrary: some teachers think a rubric with many levels is more difficult to work with, while others feel just the opposite. Deciding on the number of performance levels is often a case of what is familiar, what best fits your instructional situation, or what meets the demands of your reporting situation. The number of levels also depends on the kind of information you are seeking at a particular time. The information obtained from rubrics focused on converting student work into numerical scores is quite different from that provided by rubrics or sorting processes focused on diagnosing and describing the level of student work.

In the Creating Conversations project study group session I noticed an inconsistency in the way in which a criterion was being applied to the Water Needed for Camping problem. (Inconsistency is a very common misstep and becomes less problematic with practice and good reflective discussions.) Consequently, I knew the project study group needed to spend more time articulating criteria with which to guide the sorting process as well as practicing the sorting process itself. As we did so, I wanted to reemphasize ideas that had emerged in the Creating Conversations session:

- Understanding the problem firsthand is essential if you are to form criteria that reflect the key mathematical concepts and skills in the problem. Without an in-depth understanding of what a problem requires of students, our judgments about student work can be superficial or inconsistent.
- Teaching experiences yield common insights, but those insights remain hidden without opportunities for professional conversations.
- The sorting process reveals insights into one's own teaching practice as well as into student learning.

Session Particulars

Goals

1. Continue to practice the sorting process as a means of refining your knowledge of how to set criteria.
2. Achieve greater consistency between the identified criteria and the criteria actually applied to student work.

Sequence of activities

1. Analyze the Paper Recycling problem, which includes a bar graph (see Figure 2–1) and a line graph (see Figure 2–2). What are the students (fourth and fifth graders in this case) being asked to do? What abilities are the students being asked to demonstrate?
2. Identify performance criteria by which to evaluate the student work.
3. Sort the student work provided at the end of this book into three or four piles representing similar student performances when judged against these criteria.
4. Reflect on the criteria-setting and sorting processes.

Suggestions

Before reading the study group discussion, spend time analyzing the Paper Recycling problem. Doing the problem yourself helps you formulate your answers as well as discover complexities in the problem that might be troublesome to your students.

Now, based on your analysis, build a descriptive sentence that states in simple language a single performance criterion for judging the abilities and skills needed to complete this problem successfully.

Grades 4 and/or 5

The school district of Fairfield began a paper recycling program during the 1994–95 school year.

Using the graph below create a story about the paper recycling program in the Fairfield School District. Be sure to label the graph. Make sure that you explain your thinking in the story.

You may want to consider the following information:

- During the winter holiday season, the school district closed for two weeks.
- Due to a strike at the paper mill, the school district's January paper order did not arrive until February.
- During March, the schools were closed for three days because of snow and ice storms.
- Activities at the school in anticipation of April's Earth Day Celebration made everyone even more aware of recycling efforts in the schools.

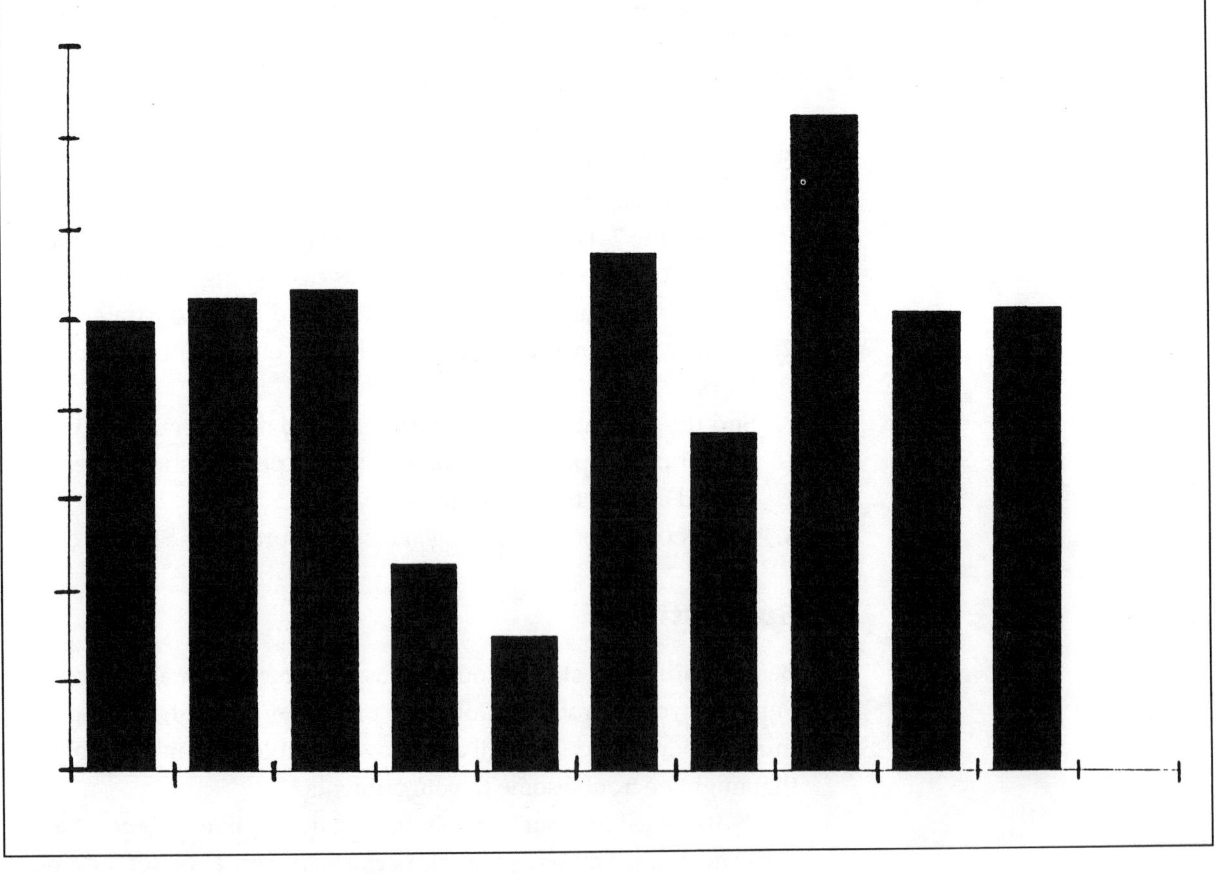

FIGURE 2–1. Paper Recycling problem bar graph.

Grades 4 and/or 5

The school district of Fairfield began a paper recycling program during the 1994–95 school year.

Using the graph below create a story about the paper recycling program in the Fairfield School District. Be sure to label the graph. Make sure that you explain your thinking in the story.

You may want to consider the following information:

- During the winter holiday season, the school district closed for two weeks.
- Due to a strike at the paper mill, the school district's January paper order did not arrive until February.
- During March, the schools were closed for three days because of snow and ice storms.
- Activities at the school in anticipation of April's Earth Day Celebration made everyone even more aware of recycling efforts in the schools.

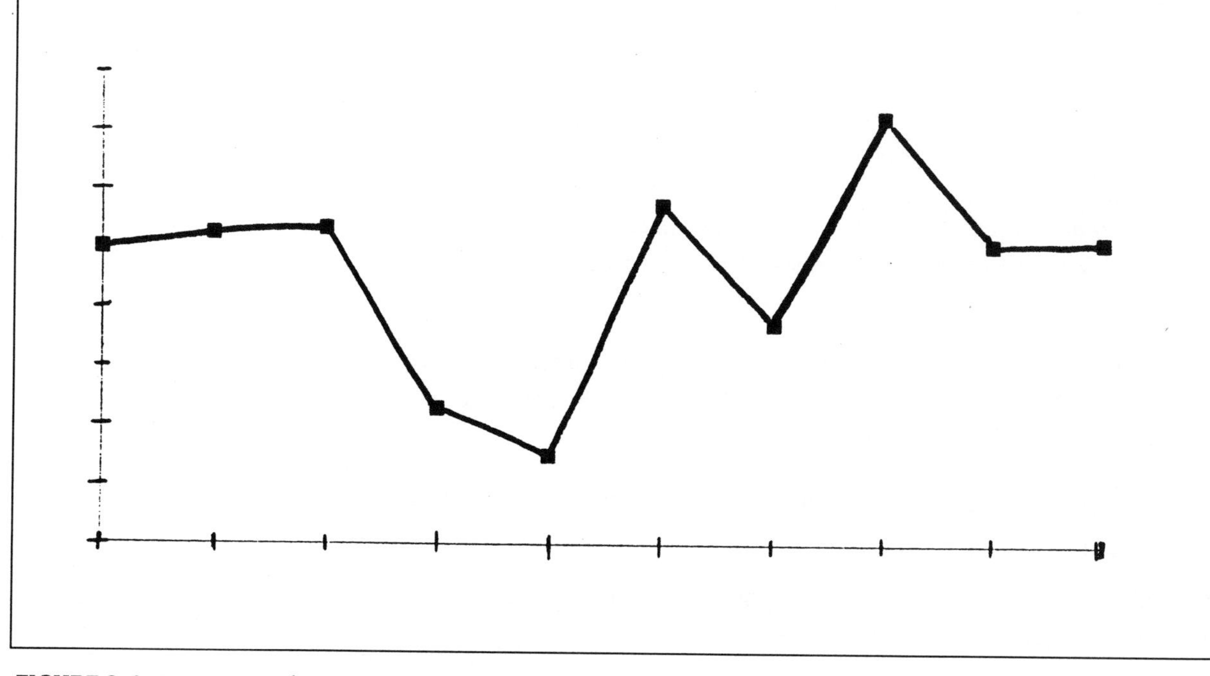

FIGURE 2–2. Paper Recycling problem line graph.

For example: *The student demonstrates an ability to communicate the mathematical information contained in the graph in a meaningful way.*

Once you've identified a performance criterion, make three or four columns on a sheet of paper to represent three or four sorting categories. As you make a judgment on a student's work, write the name of the student in the appropriate column, along with any comments about that student's work.

If you are working with this book in a study group, I recommend separating into smaller working groups. First, it's easier to have a conversation among three or four people than it is to reach consensus in a discussion held by a larger group. Second, it's important for each person involved to give her or his perspective on the student work and the application of the performance criterion to that work. The reliability of the sorting process increases as meaning is created based on shared perspectives.

When you've finished, read the project study group's conversation below. As you read, reflect on the similarities and differences between Bonnie, Gail, and Maxine's process and your own. Discuss those similarities and differences if possible.

Project Study Group Discussion

The study group separated into three smaller working groups of three or four on the basis of grade level: first and second, third, and fourth and fifth. The student work Bonnie, Gail, and Maxine are talking about (see the work samples for Session 2, beginning on p. 121) came from students in grades four and five. However, in their sorting they are not concerned about whether a particular sample comes from a fourth grader or a fifth grader. Looking at student work this way, without worrying about grade level, reminds us that grade-level expectations can be prejudicial. What should be of interest in any sorting process is the work of the individual student. Many fourth graders demonstrate an ability to construct an answer equal in quality to that of fifth graders. In fact, some fourth graders construct a better answer than some fifth graders.

Bonnie: Do you want to try three piles first? Weak, medium, and strong? Maybe pile one is the weak pile. So we're looking to see if they labeled the graph in a logical way?

Gail: That's fine, let's go with three, and pile one is low and pile three is high.

Maxine: Umm, that's fine. And, right, we're looking to see if they labeled the graph in a logical way.

Gail: Cathy's is the first one.

Maxine: There is a problem here. This doesn't look like something she would do. I really don't understand what she was trying to do.

Bonnie: Let's look at these as if we didn't know the students. What if you didn't know her? What would you think?

Maxine: I would think she needs some kind of intervention. I would pull that one from the pile. I don't think we should even include her.

Gail: Now the next student, Kate, did a nice long story. Kate has the graph starting in September. That works, doesn't it? And she has a scale that goes up, but we don't know what the scale's purpose is.

Bonnie: She says that the average of the program was five thousand sheets. Does that make sense, the idea of average?

Maxine: It does make sense. You can see the way she has labeled the graph. The only thing Kate didn't do that I think she should have is to label the vertical axis in an appropriate way. You don't know it is sheets until you read her explanation.

Gail: That next one is Sam's. He has done labeling.

Bonnie: He skipped October. That might be a little slip. Does everything else fall into place, or did he do it on purpose?

Maxine: No, I don't think that was a slip. I think Sam made a mistake.

Bonnie: Does it work, then? He has December as pretty high. The information says that during the holiday season they close for two weeks, so that interpretation doesn't quite fit. It may be something he overlooked. And then February is very low.

Maxine: They didn't get the paper until February. February is a short month, and they have a week's vacation. February could very well be the lowest month; that's logical.

Bonnie: But in terms of recycling?

Gail: Should we read Sam's description?

Bonnie: I read it. He said it did not have any pattern, but he's looking for patterns. He went up, down, down, up. The most paper recycled was in May. The lowest was in February with one hundred and fifty piles of paper. Let's put him in pile two—the middle pile.

Gail: Yes, that's what I thought.

Gail: Robbie has labeled the side of her graph "boxes of paper." It starts in September. She's got an introductory statement.

Maxine: Are you evaluating her writing?

Gail: Well, I'm combining it with what she wrote. She labeled it, "Paper Drive of 1994/95." She has months, boxes of paper, and then she concludes with, "This is the report of the paper drive."

Bonnie: I say a three. Darcy's graph is labeled "Absent Days." She begins in January.

Maxine: She told us why she left out July.

Gail: But April should be a biggy, right?

Bonnie: No, May should be a big month.

Maxine: I really think they should both be big, don't you? Because of the recycling efforts made in both of those months. I mean January, February, and March should be down, and those are not down, so I would say, put her in pile one.

Bonnie: Yes.

Gail: John's is next, and he did a line graph.

Maxine: Let's put him in three.

Bonnie: What are the attributes?

Maxine: Well, he labeled the graphs—correctly, I think. He stated what the project was.

Bonnie: Is there a correct answer?

Maxine: Well . . . no. He took the data given. He graphed the information and wrote about it so that everything makes sense. He interpreted the data.

Bonnie: Okay. And he kind of sums up a little bit of a pattern, or at least he's looking for a pattern. During the next few months, the amount of paper is pretty even. At the end of the year, it leveled off.

Gail: I think he just reiterates the same information but provides no explanation.

Maxine: Right. But why does he say, "In December the program rocket-lifted?" From thirteen to fifty-nine, from November to December. It just doesn't make much sense.

Gail: I say it is more of a two.

Bonnie: I can't decide between calling his response a one or a two. Maybe we should divide the second pile, for a total of four piles. Is that what we should do?

Gail: Yes, let's put his in pile two.

Bonnie: Cassandra has labeled the one axis. She doesn't say anything about it really. What does she mean by "rose up to thirty-six, dropped down to thirty-five"?

Gail: But she does explain the pattern.

Bonnie: Right, the pattern fits.

Maxine: Yes, that's true. So a two? *[The others agree]*

Bonnie: Daniel's is next. He skipped some months.

Maxine: Daniel didn't label some months on his graph. Why don't we have a December and January? This explanation is terrible. He writes, "The recycling was steady." The pattern isn't that way at all.

Bonnie: Is this between a one and a two? It's not strong enough to be a two.

Gail: I don't think it's as far off as a one.

Maxine: I'd say it's a two, and we have four piles.

Bonnie: Oh, really? Oh, right.

Gail: Yes.

Bonnie: Kendra's is next, and what I like about hers it that she really tries to develop a story. She gives reasons for developing a recycling program. I guess it goes beyond what is required in terms of labeling the graph, but that makes it a better piece of work for me.

Maxine: She has the affective side of the recycling program written into her story.

Gail: I agree, but how is the labeling?

Bonnie: Well, her label for pounds is not quite right, but that isn't a real problem, I don't think. I would put this one in the fourth pile.

Maxine: But is our sorting criterion how accurately the graph was labeled or the ability to create a story? Though I like the story, and she shows an understanding of the whole thing. She really has gotten into her story.

Bonnie: The papers on which the students make more of a story of it change my perspective on how these have been sorted.

Gail: But we didn't start with the idea of looking at the story part, did we?

Bonnie: Let's regroup a minute.

Reflections

As with many open-ended mathematical tasks, the Paper Recycling problem, at first glance, appears to be a simple problem, but it also yields interesting, multidimensional student data. When I asked the whole study group what this task requires of children, the responses I received were varied:

- "It's asking them to demonstrate an understanding of the purpose of the graph."
- "I think it's interpreting data."
- "I think it is about labeling a graph, doing it appropriately."

The variety of responses reflects the range of student performance information available in this one problem.

I then suggested that the study group members think about the role of the story in this problem: Why was the opportunity to create a story included? What additional information on student performance does it provide for us? Here are some of their responses:

- "What they wrote should have been sequential. It should have had a beginning, a purpose, with an outcome that matched the graph."
- "It wasn't the actual story that was important. It was that if you followed it, it made sense according to the ups and downs of the graph and according to the way they labeled it."
- "I guess our expectations would be to see if they were able to communicate information that shows relationships to the recycling efforts."

How do you think the story-building portion of the Paper Recycling problem adds to your knowledge of a student's mathematical ability? How would you answer the questions I posed? Asking what abilities are required of students to solve a problem or what the problem contributes to our knowledge of student abilities focuses our attention on identifying performance criteria. Such questions are an excellent starting point when you do not want to use a predefined rubric to judge student performance, when you want more detailed diagnostic information about your students.

Because open-ended problems allow students to demonstrate a range of abilities, criteria for evaluating students must be carefully stated and consistently adhered to throughout any sorting process. Reviewing Bonnie, Maxine, and Gail's conversation, we find several occasions when they were not as consistent and clear in defining and applying performance criteria as they might have been. Let's revisit portions of their conversation that demonstrate this gap between saying and doing.

Initially Bonnie asks whether "they labeled the graph in a logical way" is the criterion on which they will judge student work. Gail and Maxine both say fine, but the criterion is not discussed: for example, *logical* meaning what would be expected, or *logical* meaning it makes sense given the information provided by the student? Rather, all three teachers dive into the sorting task, appearing to stick to Bonnie's suggested criterion about labeling the graph in a logical way.

Bonnie should have defined what she meant by "in a logical way" or Maxine and Gail should have asked her to clarify the term. Without that clarification, each teacher interpreted the phrase based on her own particular experiences. My own understanding of "in a logical way" is that the graph will be labeled in a manner that is appropriate for a bar graph, the horizontal and vertical axes both labeled in units of measure that make sense given the information in the problem. I would also assume that the graph itself would be titled in a way that helps me understand the information contained in it. However, this may not be what Bonnie meant, and Maxine and Gail no doubt had their own interpretations. Taking the time to create clear descriptive statements that articulate performance criteria or performance indicators lessens the risk of inferring meanings that were not intended and that will lead to some confusion in the sorting process, as it did for Bonnie, Gail, and Maxine:

> **Gail:** Robbie has labeled the side of her graph "boxes of paper." It starts in September. She's got an introductory statement.
>
> **Maxine:** Are you evaluating her writing?
>
> **Gail:** Well, I'm combining it with what she wrote. She labeled it, "Paper Drive of 1994/95." She has months, boxes of paper, and then she concludes with, "This is the report of the paper drive."

A criterion can become elastic, expanding and shrinking from its intended meaning. In this case, the criterion was expanded to include the story the student created around the graph. Neither Maxine nor Bonnie pursued Gail's expanded interpretation, and consequently their saying and doing became inconsistent, steadily increasing as they began to look not only for aspects of labeling but for ways in which the students' writing reflected their understanding of the information contained in the graph. Note their responses to Kendra's work:

Bonnie: Kendra's is next, and what I like about hers is that she really tries to develop a story. She gives reasons for developing a recycling program. I guess it goes beyond what is required in terms of labeling the graph, but that makes it a better piece of work for me.

Maxine: She has the affective side of the recycling program written into her story.

Gail: I agree, but how is the labeling?

Bonnie: Well, her label for pounds is not quite right, but that isn't a real problem, I don't think. . . .

Eventually Maxine, Gail, and Bonnie realize their inconsistency between saying and doing. Maxine says: "But is our sorting criterion how accurately the graph was labeled or the ability to create a story? Though I like the story, and she shows an understanding of the whole thing. She really has gotten into her story." Bonnie continues, "The papers on which the students make more of a story of it change my perspective on how these have been sorted." Finally, Gail states the inconsistency outright: "But we didn't start with the idea of looking at the story part, did we?"

A good rule of thumb when sorting student work, even if you are using a predefined rubric, is to ask yourself, "What is the criterion I said I would use to evaluate this student work and am I, in fact, consistently using that criterion?" This question becomes even more important if you are using more than one criterion. In the beginning you may want to ask it after every third or fourth student product you assess. As you gain experience, you may find that you need to ask this question less frequently, though you should never stop asking it altogether. The benefit of sorting student work with two or three of your colleagues is your combined ability to keep the sorting process consistent between saying and doing.

While we want to strive for consistency when we judge student work, our inability to achieve it may mean we need to adjust the wording of our criteria. While we are refining our criteria, we must not be afraid to make adjustments so that we can do a better job evaluating student work. We need to make sure that all our judgments use the same criteria, and that if we do make changes midstream, we reevaluate the judgments we've already made.

More Ideas for You or Your Study Group

1. Collect samples of your students' work on an open-ended mathematics problem. This work can be from the same grade level or adjacent grades, like third and fourth or first and second. Ran-

domly select a subgroup of student samples, perhaps ten or twelve.

2. Complete the problem yourself.
3. What did you discover about the problem as a result of completing it? What mathematical abilities or general thinking abilities were necessary? (Talk these questions through with a partner if you are part of a study group.)
4. Make a list of the mathematical and critical thinking abilities you identified. Select one or two abilities that can be broken down into descriptive statements to be used as evaluation criteria for the sorting process.
5. Reflect on your understanding of what is meant by the criteria. Remember Bonnie's "in a logical way" reference to labeling the graph and make sure you are clear (and have consensus, if you are working with a group of colleagues) about the meaning of the criteria you will be using to sort your student work.
6. Using the agreed-on criterion or criteria, sort the student work samples into three or four piles that suggest high, medium, and low or beginning, developing, and advanced achievement. (If you are working in a group, do this in two- or three-person subgroups.) Record why you sorted a student work sample as you did. The number of piles or the language used to label those piles is not important. Use whatever conventions are agreeable to you and your colleagues. Spend at least thirty minutes sorting the work and thinking or talking about the *why* behind your judgments. If you take less time than that to sort ten to twelve work samples, you are probably not examining them deeply enough. You will make surface judgments without sufficiently exploring how to apply the criteria to the student product. As you become more familiar with sorting student work, your increasing knowledge will allow you to work more effectively in less time. In the initial stages, however, invest the time in the process; it is well worth it.
7. Think about your observations. (If you are working as a group, share them with the other members.) Possible questions to get started include: What stands out for you as you sorted the student work? What was most difficult about the sorting process? Were there any surprises for you in this process? Things you didn't expect? Did the criteria you used to evaluate this work allow you to discriminate among the work samples so you could sort responses into the different piles?
8. Focus part of the debriefing on the "saying and doing" concept of sorting. What was your experience in stating the criteria and then trying to apply those criteria to the student work consistently?

SESSION 3

Looking at Student Work Across Time

Background Information

One of the most complex challenges facing teachers today is how to "capture" children's progressive development across time in ways that are meaningful to three very important audiences: the teacher, the student, and the student's parents. The study group conversations on which this book is based were often limited to student work that reflected children's mathematical understanding at a particular point in time. The next logical step in developing judgment about a child's progressive understanding of mathematical ideas is to compare work done by that child at several points during the school year using common assessment criteria.

Viewing learning as an unfolding process and alternative assessment as a way to document that process is key to understanding the connection between learning and assessment. Alternative assessment, like learning theory from a constructivist perspective, assumes there is no endpoint or singular destination to knowing and understanding a mathematical idea, such as symmetry or spatial relationships. The potential for knowing more about a concept, understanding with greater depth, making more connections, and gaining greater clarity is ever present.

Without ongoing classroom assessment it is exceedingly difficult to answer such questions as *Does this child understand data and how to use them?* and *Does this child understand how to transfer his problem-solving experiences from one mathematical situation to another?* It is difficult to provide confirming evidence of whether a child knows how to reason mathematically or how to apply her knowledge of numbers unless she has been given many opportunities to demonstrate aspects of reasoning and a knowledge of numbers in a variety of ways.

The process of "knowing" mathematics is at the heart of the NCTM standards. The curriculum and evaluation standards list different ways of knowing mathematics, from kindergarten through twelfth grade: knowing how to make mathematical connections; knowing about numbers, number relationships, and number theory; knowing how to recognize and use mathematical patterns and relationships. Helping students become proficient in these and the other identified standards or outcomes requires an approach to evaluation that assumes that learning is an ongoing process through which student understanding is constructed.

Children gradually construct key mathematical ideas such as numbers, number operations, numerical relationships, patterns, and mathematical problem solving. Key to this process is ongoing feedback. Feedback from teachers and from other students serves as a mirror, reflecting back to the student the progressive nature of his understanding.

Educational researchers Patricia Campbell and Martin Johnson (1995) suggest that the maturation of mathematical understanding is best characterized as a process, not as an incident. They note that mathematical understanding "takes time; it does not happen all at once. Further, children do not develop mathematical maturity at the same rate. . . . Children also need sufficient time to reflect and think, with the expectation that children will express and justify that thinking" (pp. 35, 37).

Traditional school practices do not always support the ongoing, progressive nature of learning. Schools have long engaged in practices that break up the learning process into standardized time units: semesters, quarters, terms, six-week marking periods. These units, or time markers, frequently signal an artificial end to studying a concept, because they make it difficult to foster the connections among topics over the school year.

The organization of commercial textbooks also influences when and how student progress is measured. Textbooks are traditionally divided into chapters and units, so the times set aside to measure student progress have quite naturally been at the end of a chapter or a major unit. The linear presentation of content in commercial textbooks also directs evaluation toward quantity: How much of the material presented has the student mastered? Is this level of mastery sufficient for her grade level?

However, when assessment becomes a part of everyday instruction, questions asked about student progress are noticeably different—they more easily focus on progressive understanding rather than on the mastery of predefined content. In an alternative assessment model, it is unlikely that student learning will be represented only by achievement scores; there will also be descriptive evidence, based on common criteria, of the ways in which students demonstrated their knowledge of targeted mathematical concepts. Likewise, there should be different sources of information from which to construct a pattern of learning and different ways in which children are asked to interact with an assessment—work on a project, compose a written response, provide short answers, or build a portfolio.

At the heart and soul of alternative assessment strategies is a desire to know about the progressive nature of a child's understanding, not just his factual knowledge. An alternative assessment perspective assumes that student understanding is more than the sum of discrete skills and facts accumulated so far: in order to be useful knowledge it has to be "put to use." While it can be important to memorize definitions, math facts, and algorithms, the complete measure of a student's understanding cannot be taken until those definitions and facts are applied in situations that are meaningful to students. Rieneke Zessoules and Howard Gardner (1991) comment:

> Teachers must ask themselves: "How long does it take to nurture students' habits of mind? When is it appropriate to assess students' work? What kinds of qualities would I look for initially? How might those dimensions change over time? How do I document the broadening of these skills and abilities?"
>
> The answers to these questions do not come easily. On a practical level, things happen slowly in an assessment culture. Students cannot be marched quickly through the curriculum, because it is not composed of a series of activities that yield discrete products, but rather a set of opportunities that encourage complex habits of mind, ways of working, and processes of learning. These processes mature and develop in an ongoing, if bumpy, way. (p. 62)

Session Particulars

Goals

1. To identify conceptual links between assessment tasks given at different points in a school year.
2. To apply a given set of assessment criteria to work second graders produced at two points in the school year.
3. To talk about the progressive development of a mathematical concept and to practice making descriptive statements that focus on progressive student development.

Sequence of activities

1. Examine the two mathematical tasks for second graders shown in Figures 3–1 and 3–2: how are these tasks similar in terms of the knowledge and skills necessary to complete the task?

List some ways that these graphs are alike.

Tell a classmate why you think so.

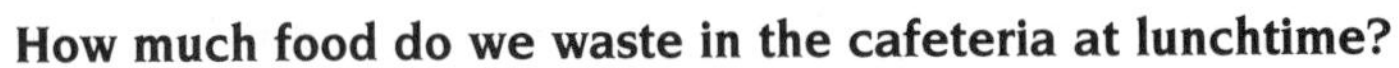

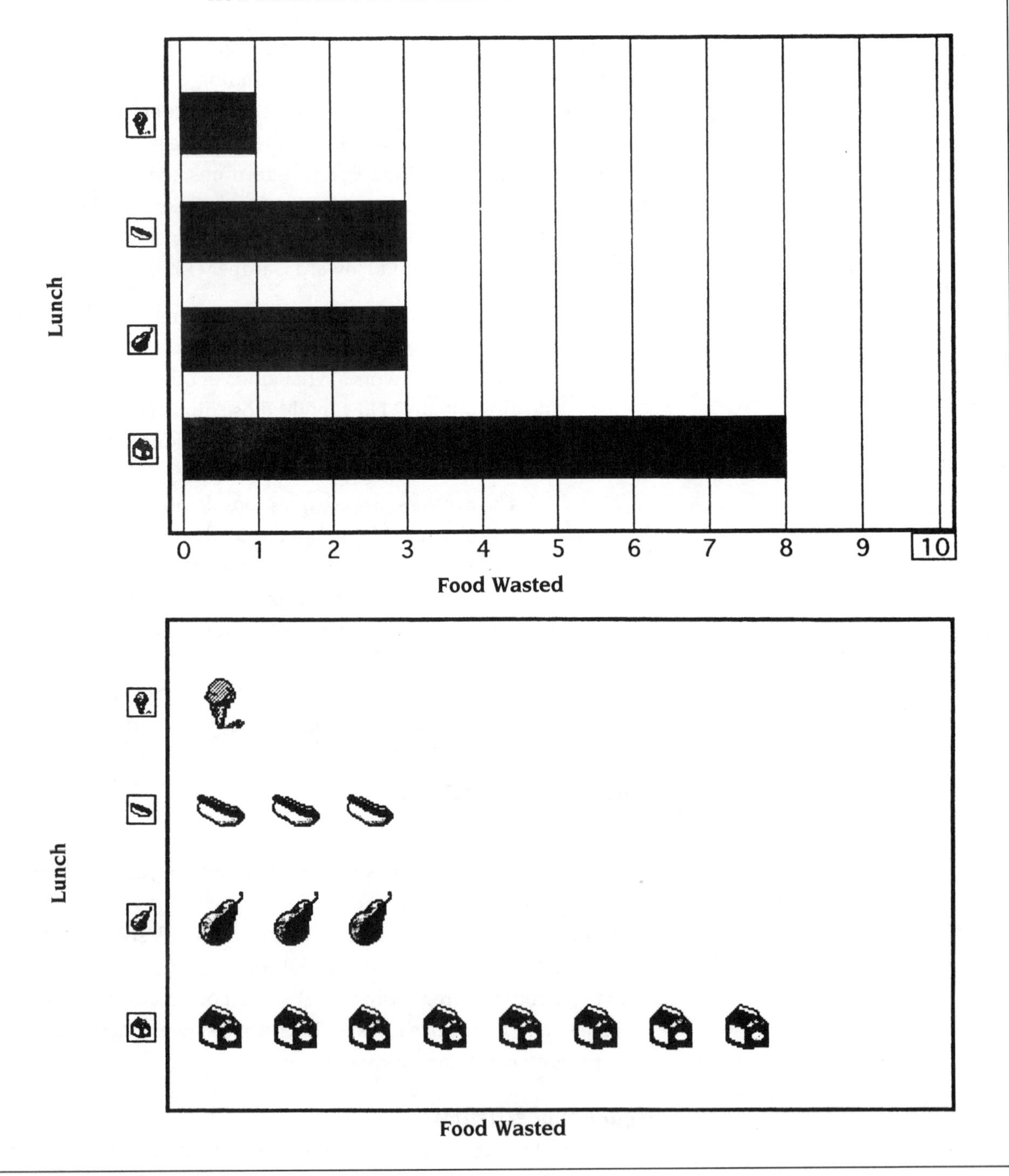

FIGURE 3–1. Problem for first and second graders in November.

Tell all you can about these two graphs.

How are they alike?

How are they different?

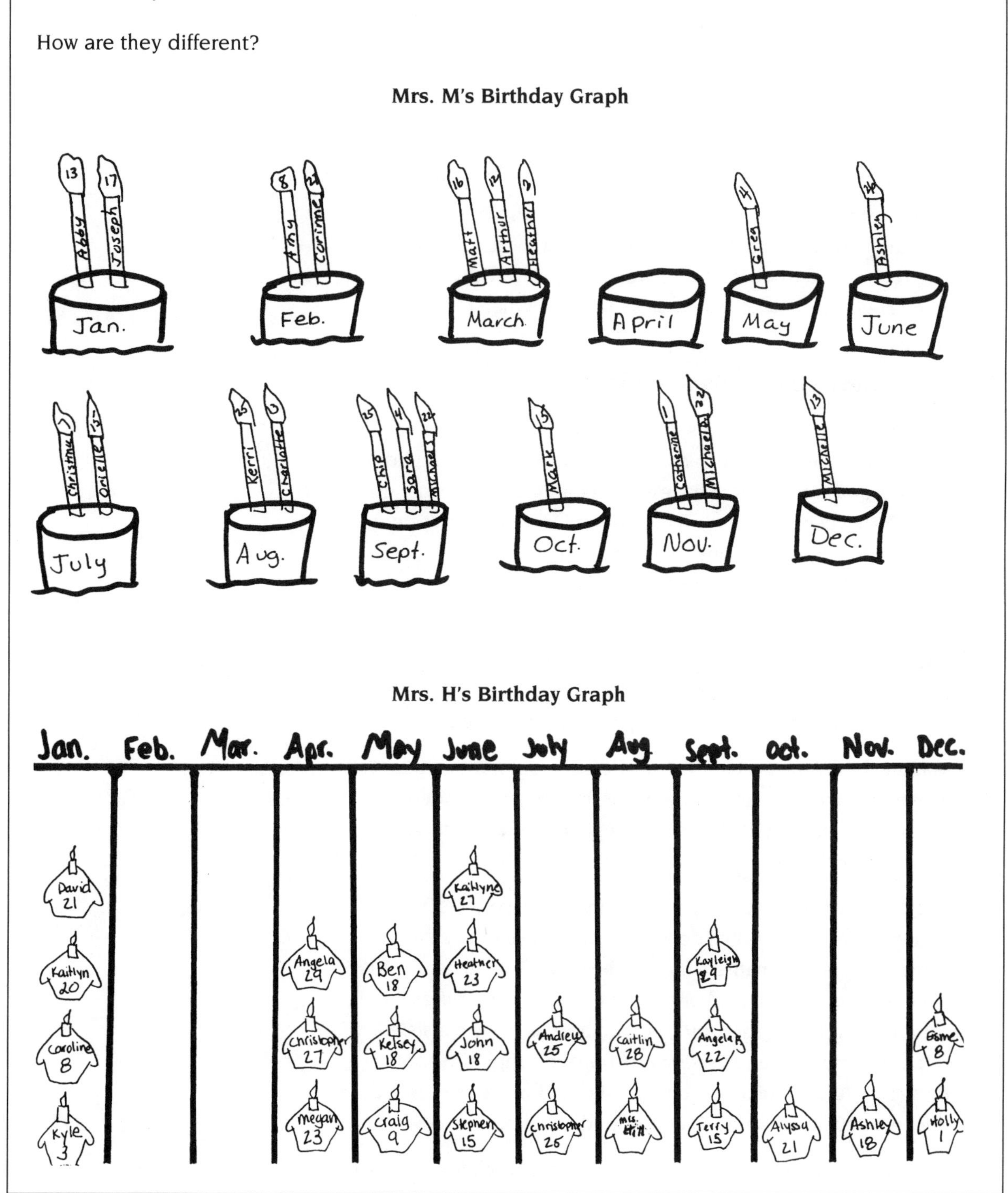

FIGURE 3–2. Problem for first and second graders in March.

2. Review and discuss work samples of one student completed at two points in the school year in terms of the similarities you have identified.
3. Prepare notes to be included in the files of this child in preparation for an upcoming parent-teacher conference.
4. Reflect on the process.

Suggestions

Examine the two mathematical activities about graphs (Figures 3–1 and 3–2). How are these tasks similar in terms of the mathematical abilities they require of students? Or, more specifically, how are these tasks similar in terms of what they are asking of a second grader?

Next, review and discuss Andy's work samples at the end of this book (pp. 141–43) and make some notes for Andy's file.

Project Study Group Discussion

JM: Throughout our conversations about student work, I've often heard comments such as, I am really seeing this child differently, or, I would never have expected this child to come up with this level of response. In our attempt to connect classroom instruction and assessment, you seem to have come to know some of your students in a different way. Would you agree with that?

Kate: Yes. Doing this has helped me become more discerning about what the children are doing mathematically—all the time, not just in that one isolated testing piece. It's very helpful.

Maxine: I think just expanding the definition of mathematics to go beyond symbols, to include words and the open-ended discussions we have made part of our lessons, invites participation from more children.

Lucy: These discussions have helped me in my conversations with parents, because I've developed new ways to talk about a child's progress in mathematics. I'd been stuck on general sorts of comments: he is doing better with conserving numbers, her computation is more accurate, her accuracy is more consistent. But until now I haven't used instructional examples to support those sorts of comments. I guess I just didn't make the connection; we depended so much on the grade book.

Sharon: Parents always want to know how their child is doing in comparison with the other children in the class. Even if you give them more descriptive information, it still seems to come back to knowing what that means in terms of the other students. It's hard to get out of that mind-set.

Vicki: That's probably true, but maybe we haven't been able to provide parents with other models of student progress.

JM: What Vicki is suggesting makes sense. I'm not sure we have developed good ways of describing children's progress other than comparing one set of children with another. However, our work today will be an opportunity to think about new ways to talk about or describe student progress. So let's turn to Andy's work. First, let's look at the assessment tasks themselves. In order for us to have a meaningful conversation about Andy's growth in understanding mathematical concepts, we need to find the similarities between the two assessment tasks. Once we've found those similarities, we can identify common assessment criteria and document Andy's progress by comparing his work in November with his work in March. Why don't we start by focusing on this question: How are these two tasks similar in terms of what they are asking of second graders?

Gail: Both tasks are asking students to make comparisons. Both problems have two graphs that display information in different ways. The students are going to have to be able to interpret each graph and then compare one with the other.

Maxine: That's right, both tasks ask students to make comparisons and then interpret or explain those comparisons.

Francine: Both tasks assume that students are able to communicate about the comparisons being made, that they know how to write comparatively. They need to recognize the comparisons, but they must also have the mathematical language to write about them.

Bonnie: Let's not forget that a child would have to have some understanding of the different kinds of graphs and how graphs represent information, frequently numerical information.

Kate: I need to understand something about what we're doing. I'm fine with looking for ways in which these two problems are the same; what bothers me is that Andy isn't repeating the same problem. Is it fair to make statements about how a student has progressed or hasn't progressed if the problem isn't the same problem? The problem given to the second graders in March asks the students to deal with similarities and differences; the problem in November just deals with similarities.

JM: Historically, educational research has been influenced by the experimental model—the researcher wanted to know the effect of one situation on another. In that situation, the item or characteristic or trait being measured is frequently held constant: you do everything possible to assure yourself that what you are trying to measure or evaluate, often called the independent variable, remains the same. The *dependent* variables are changed. For example, a research question might be whether class size has any effect on children's math scores. The researcher studies a number of classrooms of varying sizes and the math scores of the children in those classrooms. The math achievement score is the independent variable, and class size is the dependent variable. The question we are asking here is different. We want to discover the pattern in a child's understanding of a mathematical idea or concept; we're not as concerned with what might have influenced the pattern as we are with identifying it. So our job is to look at how children apply similar mathematical concepts at different points in the school year, in order to find evidence or clues of their growing understanding about those concepts. That's a different kind of a job than isolating what things may have contributed to this understanding.

Kate: Okay, this is making sense. The issue isn't how well Andy scores on a specific problem but how his understanding of graphs is changing.

Francine: That's an interesting point. Often when we talk about children's growth, we are really comparing scores or grades that don't reflect the same math content or ideas. I know some topics that are introduced in the fall may not be addressed again until the following year.

Vicki: Maybe when we look at a child's growth in math in a given year, we are talking about growth at a general level—like Andy is doing better in math now than he did during the previous six-week period. What we haven't done is figure out how to make the curriculum and assessment work together, how to follow these bigger ideas, like talking about numbers or interpreting data, across an entire school year.

Lucy: I think that's the coverage issue: how can we cover all the topics we are supposed to cover and still find the connecting ideas?

JM: Lucy, your comment demonstrates again just how closely related curriculum, instruction, and assessment can and should be. For example, let's go back to the similarities we were identifying between these two problems in terms of the mathematical knowledge required to complete each one successfully. According to my notes, we said the student abilities necessary for both the November and the March problem were, one, being able to write comparative descriptions, to communicate; two, being able to think comparatively; three, being able to understand and recognize different kinds of graphs; four, being able to recognize similarities; and five, being able to interpret mathematical information presented in a graph. There may be other common abilities, but these five give us a place to start our conversations about finding evidence of growth. What I would like us to do now is look at Andy's work in November and in March, zeroing in on one of the mathematical abilities common to each problem. Let's see whether we can identify evidence that Andy has grown in this ability between November and March. Which ability should we use?

Sharon: I'm interested in the ability to communicate comparisons—in this instance, to write comparatively.

Jo-Anne: I think that's a good choice, because even though this is a second-grade problem, this ability goes across grade levels.

JM: Okay, let's look at Andy's work with just that criterion in mind.

Sharon: Just the change in handwriting is stunning!

Maxine: Yes . . . though maybe that could be misleading. His ability to communicate comparisons may not be similar to his growth in handwriting ability. Though it is hard not to be positively influenced by that growth in penmanship!

Jo-Anne: In November, Andy indicated he thought the graphs were the same in many ways, but he didn't write down what those ways were. In March, he did write in more detail about the comparison he was making.

Lucy: Yes, he did. But what I noticed is that his work in March focuses only on differences between the two birthday graphs. He does not identify ways in which the graphs or the information contained in the graphs are alike.

Kate: But I think there is evidence of growth in his ability to write in more detail about the information in the graphs. In November he suggests he knows there are similarities, but he does not provide detail. In March, he identifies differences between the two graphs and does provide some descriptive detail. So I think that's one piece of evidence we would use to describe his growth related to this criterion.

Gail: Providing more descriptive information in a comparative way is noticeable, but I disagree with Lucy a bit. Andy does refer to similarities between the two classes. But he seems to concentrate on comparing the number of birthdays in each class, and all his descriptions are lumped together, like a run-on sentence. The similarities and differences don't stand out. For example, "In Aug. there is two people at Winslow and two people at my class too."

Lucy: Gail, you're right. I read it too hastily.

Bonnie: I noticed his sentence about there being twenty people in his class and twenty-four people at Winslow. He took a step toward making comparisons between the two graphs using information beyond the obvious. He had to total the number of birthdays in each class and compare those totals. Though his addition wasn't correct—there are twenty-five students in the Winslow class, not twenty-four—he did take that next step.

JM: If you were writing a note for Andy's folder about his understanding of comparisons and his ability to communicate his understanding of comparisons based on these two pieces of work, what would you write?

Bonnie: I would want to note his growth in providing comparative detail. In March he could really write about what he was seeing in the two graphs.

Vicki: I would mention that he needs more practice in organizing his comparative thoughts. I would want him to spend time practicing how to organize the way he presents his comparisons. Maybe he should start by making two columns on a page, one labeled differences and the other similarities, or he might organize the information using Venn diagrams. Then he could match his observations to either of these organizing strategies. Once I was sure he is clear about both concepts, similarities and differences, maybe then I would encourage him to practice writing sentences that have comparing and contrasting qualities.

Lucy: I would mention that I would try to reinforce his growth in these concepts by having him do similar sorts of practice steps in other subjects. My sense is that he is trying to classify data into categories, but he is struggling with how to express or communicate the attributes of those categories.

Francine: That kind of information would be very helpful to parents, since these things could be practiced at home as well.

Jo-Anne: To go back to what we talked about earlier, looking at

Andy's work in this way, finding his unique pattern of growth from one point in the school year to another, was helpful to me. Too often we compare students with one another rather than describe individual growth. I've been thinking about portfolio assessment. I know I want to be able to capture individual growth, but I haven't been quite sure how to go about doing it. I am going to spend time identifying mathematical concepts that the students can work on throughout an entire school year.

Reflections

Focusing on one student's work at different points during the school year is valuable. Looking at a number of students doing one task and looking at one student doing a number of tasks that require similar mathematical abilities provide different kinds of information. Both kinds are necessary, and each is complementary to the other.

Gathering evidence of a student's performance at a particular point in time reveals progress to date on key concepts. Teachers can use this information to answer questions such as:

- Is there a pattern in the errors the student is making in this activity? What is the evidence for this pattern? How can it be conveyed to the student?
- Based on the outcome of this assessment, what should the next instructional steps be?
- Is there information contained in the results of this assessment that should be included in the student's folder?

When teachers gather evidence of a student's performance at multiple points, they obtain greater insight into the progressive development of that student's understanding of mathematical concepts. Focusing on these patterns allows them to answer a different set of questions:

- How does this evidence help put together a descriptive pattern of growth around a mathematical concept or big idea?
- Is this pattern revealed in other subject areas?
- How can this pattern be shared with the student? Would asking him or her to assess these same examples of work help?
- Is there evidence of a student misconception that might be helping to shape the pattern we are seeing?

Both methods of classroom assessment clearly have a place for student self-assessment. It would have been helpful to know Andy's thoughts about his work in November compared with his work in March. Students need the opportunity to look at their work in a thoughtful manner, and need guidance on how to identify evidence of growth in their understanding of key concepts as well as areas for further practice.

Frequently, when children look at school papers from previous years or from an earlier time in the same school year they will say, somewhat incredulously, *Boy, I sure didn't know much then!* I have noticed my own son, now in high school, studying work he completed as a middle school student and shaking his head at what he didn't know, rather than celebrating the things he obviously knew quite well.

Alternative assessment can be a powerful tool in broadening everyone's understanding of the multiple ways in which student learning can be documented. Students themselves, through thoughtful self-assessment strategies, can be an integral part of this documentation process. If conversations among teachers, among students and teachers, and among parents and teachers can occur in an atmosphere that presupposes the progressive nature of learning, complete with steps that sometimes go backward before they can go forward again, then evaluation can be of real service to classroom instruction and the curricular materials used to support that instruction. Greater emphasis on creating conversations about the progressive nature of learning can better support the role of alternative assessment as a diagnostic tool for teachers and the role of the teacher as the diagnostician.

Traditionally, the questions about student evaluation have focused on *when* and *how.* When do we do it? Every week? Every six weeks? At the end of a chapter? How do we evaluate them? With a commercial test? A group project? A short quiz? An emerging question is *why.* Why are students being evaluated, to what end? A very helpful distinction can be made between assessment for purposes of accountability and evaluation or assessment for purposes of information to aid teachers', students', and parents' awareness of issues in the teaching and learning of mathematics.

Accountability will always be there, because education costs a great deal of money and taxpayers and legislators need to be able to "account for" those costs. However, classroom or building-level assessment that informs teachers, administrators, students, and parents about students' ongoing development and documentation of key curricular standards and concepts is a relatively new idea that alternative assessment has helped to introduce. If we are to continue to make the case for alternative assessment, we need to show teachers and students how to integrate it into the classroom effectively.

More Ideas for You or Your Study Group

1. Look at Andy's work from the perspective of one of the other abilities identified by the group as common to both mathematical tasks (the ability to understand and recognize different kinds of graphs, the ability to recognize similarities, or the ability to interpret mathematical information presented in a graph).

2. Identify other abilities common to both tasks and create a conversation around one of them.
3. The work of Andy's second-grade classmate Samantha is also included at the end of this book (pp. 145–47). Using the dialogue of the study group as a guide, talk about Samantha's growth from November to March in one of the identified abilities.
4. Write a note for Samantha's folder about her progress to date in mathematics based on your conversations about evidence of progress.
5. Find examples of mathematics work over time among your own students. Assess this work for evidence of progress as you did with Andy and Samantha. You might also want to talk about what changes, if any, you would make in your mathematics curriculum to enhance the opportunity to document growth in mathematical concepts.

SESSION 4

What Constitutes a Good Classroom Assessment Activity?

Issues related to designing alternative assessments are complex, and the ideas presented in this session are therefore not intended to be complete—clear, yes, but not complete. This session focuses on informal classroom assessments in which the goal is to obtain information on teaching and learning, not conduct a formal evaluation. We need to strive for precision in our language about alternative assessment; the enormity of the subject demands it. I am grateful to Elizabeth Badger for reminding me of how entangled we can and do become in our written and spoken explanations. I admire her clarity on these issues.

Background Information

Open-ended problems are an increasingly visible part of our efforts to reform mathematics, since they provide the opportunity for children to present evidence of their mathematical thinking and their progress toward essential learning outcomes. The old way to try to uncover students' mathematical thinking was to ask them to "show their work," which generally meant showing their computational steps. Today, as we work to refine our thinking about alternative assessment and its role in learning mathematics, we know that open-ended problems and investigations give students a context for presenting evidence of their mathematical thinking that goes beyond computational skills.

Open-ended problems or investigations are an excellent pathway linking instruction and assessment. These sorts of classroom activities frequently have students retell in their own words the steps taken to solve a problem or share their thinking about a problem, thereby opening a window through which to judge the progress of student learning. As a result, teachers can make decisions about topics or content ideas that need more practice and/or instruction.

The idea of ongoing, oftentimes informal, assessment in the classroom is relatively new. Historically, student assessment was a means of providing a picture of student achievement that was used to rank, sort, and rate students, schools, and school districts. Little, if any, attention was given to evaluation as a means to help teachers adjust their instruction or to help students understand their own learning. Greater understanding of the role of reflection in the learning process and the role of meaningful, contextualized problems for the learner has done a great deal to broaden the purpose of evaluation to include the needs of both students and teachers. Joan Herman has this to say about the shift in both assessment and instruction:

> Current evidence makes it clear that instruction emphasizing structured drill and practice on isolated facts and skills does students a major disservice. Insisting that students demonstrate a certain level of arithmetic mastery before being allowed to enroll in algebra and [insisting] that they learn how to write a good paragraph before tackling an essay are examples of this discrete skills approach. Such learning out of context makes it more difficult to organize and remember the information being presented. Applying taught skills later when solving real-world problems also becomes more difficult. Students who have trouble mastering decontextualized "basics" are often put in remedial classes or groups and are not given the opportunity to tackle complex meaningful tasks. (Herman et al. 1992, p. 15)

One of the primary concerns of recent work in alternative assessment, therefore, has been to create tasks that provide teachers, students, and parents with ongoing opportunities to listen for, observe, and read student understanding. These tasks help us answer the question, *How do we know that students understand something?* Certainly, we must continue to work toward high performance standards in mathematics; students need to master key mathematical ideas and concepts. However, of much greater interest is *how* students construct their understanding of mathematical ideas on the path toward achieving those high performance outcomes. Understanding and achievement are not opposing ideas. High achievement and an understanding of how learning is progressing in relation to instruction on the way to mastery are complementary goals. Improved understanding of how all students learn mathematics and greater involvement by students in how and what they are learning only strengthen the likelihood of higher student achievement.

To understand how children learn, we need to be good diagnosticians of their work. The instructional tasks we use in our classrooms for assessment purposes should give students an opportunity to demonstrate their progress toward agreed-on instructional goals in mathematics, and teachers must know how to interpret that progress and adjust their instructional goals accordingly.

Assessment tasks vary, depending on their purpose. An activity helpful in diagnosing and understanding student learning is not necessarily going to be the best one for determining the effectiveness of a school's mathematics curriculum. Being comfortable with the fact that there are different kinds of assessment supports the richness of the alternative-assessment concept and helps us understand the relationship between purpose—*why am I giving this assessment?*—and design—*how can I structure the assessment experience(s) to achieve my purpose?* For example, the purpose of an assessment task influences technical considerations such as reliability and validity. The reliability and validity of a high-stakes assessment are extremely important. Likewise, the information you want to get from an assessment task influences the way students are asked to interact with it: provide short answers, complete a project, write a report, or build a portfolio. Some assessment formats are better than others for finding out specific things about children's thinking.

On the other hand, informal classroom assessment primarily benefits the teaching and learning process. It helps us as teachers make necessary corrections to our teaching. In order to teach well, we need to know how well we are getting the requisite information, ideas, and experiences across to our students. The only way to know this is to talk to, listen to, and observe our students and set different levels of tasks for them to perform. Such tasks help students by providing them with repeated opportunities to get feedback that demonstrates how good a job they are doing.

Our study group often talked about how to design assessment tasks. Occasionally I would ask the participants to think about the qualities or characteristics of good informal classroom assessments—ongoing evidence used by the classroom teacher to make instructional decisions. How would they design instructional tasks or problems that could be used in the course of their everyday classroom mathematics to provide evidence of instructional progress? What qualities would they want those problems to have?

Session Particulars

Goal

To answer the question, *What are the characteristics of a good classroom assessment task?*

Suggestions for a study group

1. Break up into smaller groups.
2. Individually take about five minutes to think about mathematical problems or investigations you consider to be good ones for informal assessment purposes. (Establish your own criteria—why are these problems good?)
3. Have everyone in your small group share the tasks she or he identified and why. Have someone in the group write down the characteristics identified.
4. Reassemble the whole study group and have each smaller group present its listing.
5. Make a combined list of all the characteristics and note those that were identified more than once. Post this list where everyone can see it.
6. Discuss the list. You might want to spend time on the frequently identified characteristics and discuss why these seem to be common among group members.

Suggestions for the individual reader

1. Think about mathematical problems or investigations you consider to be good ones for informal assessment purposes.
2. Write down the characteristics you think make them good.
3. Identify why you think these characteristics contribute to making an activity a good one.
4. Think about the following questions: Have you identified characteristics that tend to be a regular part of your instructional activities? Do these characteristics represent the perspective of the teacher, the student, or both?

Project Study Group Discussion

JM: Think about the instructional activities you use in your mathematics classes and the problems we have been using in connection with our study group. What makes a problem or an investigation a good one for purposes of informal assessment? Can you identify characteristics of a good assessment task that provides you with information on your teaching and your students' learning?

Kate: I think an assessment task should pinpoint areas of weakness that you want to target for certain children. And you also want it to be open-ended so that it doesn't create an artificial barrier for children who can do more. Sometimes I have a feeling that certain children can do more, but the problems I have don't do much to validate that feeling.

JM: Good. Other thoughts?

Lucy: How clear the problem is.

JM: Lucy, let's establish your notion of clarity. What makes a problem clear?

Lucy: For me clarity and validity are related. I think a clear problem is one that is familiar to the children in terms of what it is asking them to do. For example, when I was doing a graphing activity with my students, I felt that because of the work we had done before it would be clear to them. But when I passed it out they all just sat there and looked at me. So I thought I must not have been very clear in my directions, that I had left something out. But it turned out that they had a difficult time understanding how they were going to write about the graphing activity. I guess it was just too different from what we had done in the past even though it was about graphing. That experience really told me something about my teaching.

Francine: I know what you mean. You're not sure. You think, was it my fault they didn't understand what to do?

Lucy: Right. So when I give an assessment activity, I want to make sure it really reflects what students know and are able to do; the assessment is valid because of that.

JM: Let me try to expand a little on Lucy's comment. You're saying that a good open-ended assessment activity is one that is similar to previous experiences students have had practicing targeted mathematical abilities in an ongoing way.

Sharon: An assessment activity should really engage a child; she needs to be interested in and proud of what she completes.

Maxine: I think my children's most difficult times are getting over the initial phase of, what is the answer you want? You can see their relief when I explain that this time not every answer has to be exactly the same. Their attitude toward attempting problems changes because they feel differently about the expectations. I want to have problems that really honor my students' thinking, and that is a shift for me.

JM: Kate, do you have any thoughts on this?

Kate: I'm just thinking about performance. I'm always looking to see where that child is. I'm more interested in a child's thinking over time, in the context of that one child, than I am in the broader developmental context of grade level or anything like that. I don't think of it so much as a performance that has connotations of competition and judgment. I want assessment to promote more reflection by the children and by me.

Vicki: I would expect a good problem to be able to tap into where the children have been, to tell me what they know as well as what they don't know.

Lucy: I think that is always really hard to evaluate. Every child comes with a different background, each one thinks differently. I don't know, it just seems really hard to measure.

Sharon: An assessment for me has to tell me my next instructional steps. And even though I want this information, there is a price to pay: I will probably have to go in multiple directions. Not all the children will need the same thing or be in the same place. In the old days, I let the books dictate the next instructional places to go.

Jo-Anne: Sharon, that just triggered a thought. For me a good task provokes a rich discussion at the end. After the children have been working on it, there is so much to talk about that new questions arise. I think the connection to assessment is that those discussions give me so much information about the kids.

Vicki: I also want a task that is user-friendly for the student. And then I guess I have to ask myself, what do I mean by user-friendly?

JM: Do you mean that it is an "approachable" problem for students? That it's in the realm of a student's experience and she has a sense of how to begin to work on it? That the student is somewhat familiar with the mathematical knowledge and procedures that are a part of it? Sometimes it's important to analyze an activity so that you have a full understanding of what is required to complete it successfully. A fundamental question to ask in the course of designing an assessment activity is, what would a child have to know to be able to complete this activity successfully? Then do the activity yourself, experience it yourself.

Vicki: I see a connection between this conversation and what my school is trying to do in redesigning our report cards. A committee has been working on designing our new report card format, and as I was meeting with them, it occurred to me how important good classroom assessment activities are in giving us evidence of how students are doing in areas like communication skills and problem solving. We still have to evaluate students in the subject areas, but these other areas require a different kind of information. I'm really struck by how what we've been doing in the study group fits in with our report card revisions.

JM: That's a good example. It's so important to have these conversations about what kind of information is most meaningful and what kinds of classroom activities will elicit that information.

Francine: If I choose to make an instructional activity an assessment activity, then I want the information to mean something to me. So what you are suggesting about the report card makes sense. We do get much more information from the kinds of open-ended problems we have been doing, but we also have to know what we are going to do with the information; otherwise it is just all so overwhelming.

Vicki: Does that mean our classroom assessment activities have to be designed so that there is a match between what we tell parents is important—in other words, what's on the report card—and the way we target our data collection in the classroom? It certainly will give parents much more information.

Lucy: I think if we could get a better handle on all of this, we wouldn't feel as if we are doing double duty so much of the time.

Bonnie: The parent piece to all of this is still problematic for me.

Gail: If I were designing an assessment activity, I would already have in mind the kinds of things I wanted to be able to observe in my children. I would know the things they have been working on, the skills they should be able to use. These things would influence my informal assessments. I wouldn't just be giving them a test and then figuring out what to do as a result of the test. I'm beginning to think about all this quite differently—I realize it makes more sense, especially to the children, because they know what they have been working on as well, what mathematical abilities they have been practicing.

Sharon: I find the first time I introduce something to the children it's sort of helter-skelter. Then we'll talk about what it is, what we're looking for, what the criteria are, so the next time they encounter something similar they'll know what's wanted and how to go about it. We always have some kind of sharing time when I tell them to look at their work and ask, *Did I do that? Does my work show that?* Little by little, they're able to incorporate more and more of the things we have agreed on as a group, the students and me, about what is important in the mathematics we are doing. I guess what I'm trying to work on right now is using classroom assessment as a way to promote reflective communication about mathematics.

JM: Sharon, how does your wanting to have more reflective communication in your mathematics lessons influence how you are thinking about assessment design?

Sharon: I want an assessment problem to encourage communication among the children and between them and me. We've talked about this before. I think the kind of problems that encourage talking and sharing are the ones that have a number of steps to them, where one step builds on another. Having a number of steps moves the kids away from being so concerned with getting the right answer; there is more opportunity to talk about how the problem was solved.

JM: So an important characteristic of a good assessment problem is that it has multiple steps, because you want to emphasize communication in your mathematics lessons?

Sharon: That's right.

Francine: I would add a caution about the way we ask the children to communicate. We need to be more specific about how they are to explain the reasoning behind what they are doing, especially in the upper grades. Sometimes our study group problems have asked students to tell or explain how they went about solving a problem and why, and their explanations have had nothing to do with mathematics, absolutely nothing. I think an assessment problem should specifically state, tell your mathematical reasons for doing it. It would limit writing to mathematics.

Jo-Anne: I would say, show me, because that kind of language reaches more learning styles. A number of kids in my class this year are really good artists, and one or two of these artists are not particularly good writers. I've found that giving them the opportunity to draw or make a model of their thinking lets me get more information. I tell my kids, you can show me any way you want to, I just have to be able to understand it. Sometimes they totally shock me with the variety of ways they represent their thinking. So I think a good assessment problem is stated in a way that is open to the range of learning styles and abilities we all have in our classes.

Kate: Words like *represent* and *show* give students a new language to deal with, don't you think? And I'm not even sure we all know what is meant by those terms, so perhaps that's an area we need to explore. I may have a picture in my head about what I mean by *represent* but it may not be the same as what other teachers have. That could be a problem.

Gail: Attention to language has to start early, in kindergarten.

JM: I'm hearing a couple of interesting ideas. The first is that a good classroom assessment task should collect evidence on abilities and mathematical content knowledge that are part of your mathematics instruction. This makes me think that the label *open-ended problem* may sometimes be misleading. Open-ended problems are not unfocused problems; rather, they are focused on obtaining evidence of student progress in identified areas of your curriculum. You want the design of that assessment task to be as helpful as possible in obtaining that evidence. Second, language is important in a good assessment task. Children with a range of learning styles and abilities need to be able to understand the language of a problem. When you use terms like *tell, show,* or *represent,* children need to understand what those terms mean.

Jo-Anne: So to make an assessment problem work, I mean really provide us with evidence of how students are doing in mathematics, we need to spend time in the classroom with our students and time outside the classroom with other teachers talking about expectations. Does that make sense?

Vicki: It does to me, and I agree. You can't just give an assessment activity to a student and expect that it will do everything you want it to. You have to spend time talking about expectations and defining terms.

Reflections

Throughout the year I kept an ongoing list of the things we identified as being characteristic of a good classroom assessment task, adding and subtracting ideas based on our monthly conversations. The list was not a focus of every meeting; sometimes I would mention it, sometimes one of the study group members would bring it up. Its being a work in progress underscored the purpose of the study group—to explore assessment through children's work, with the study group members having as much permission and responsibility to develop the content and ideas as I did. During our final meeting, we converted these characteristics into the series of questions shown in Figure 4–1.

The list is not definitive; it came from the work of this particular study group and reflects the perspectives and experiences of the study group members at that point in time. If I had another meeting with the study group to discuss this list, we would revise it based on the teachers' experiences since we last met. And our thinking and our ability to articulate that thinking would be sharpened.

While there is a growing consensus among classroom teachers as to what makes a problem rich in context and rich in terms of the amount of student thinking and application it provokes, agreement is certainly not universal. Local situations, philosophical beliefs, and curricular outcomes are very influential. Some of the characteristics in Figure 4–1 would probably be suggested by any study group; others might not be, and different characteristics raised instead. This is how collaborative professional development works.

Our study group agreed that a good classroom assessment task would provide ample opportunity for students to communicate their thinking. An interesting concern emerged over the "language" of the assessment activity. Some study group members felt the assessment would be unfair if the students did not understand the language being used to present the assessment activity. Lucy said that if the assessment activity wasn't clear to a student, then it wasn't valid for that student.

The traditional meaning of validity in the context of evaluation is different from what Lucy was suggesting. In the context of tests and measurement, a valid test item is one that measures what that test designer purports that it will measure, so that we have some degree of confidence that test results have meaning. Lucy's use of the term had little to do with the issues of psychometricians. She related validity to

- Does the task engage the students?
- Does it contribute to putting together a pattern of performance among children over time?
- Does it provoke good communication among students?
- Does it not depend on the teacher?
- Is the task clear to the students? Is the language of the task understood by the students?
- Does it pinpoint areas of student weakness in mathematics content and thinking ability that need more intensive practice? How?
- Do the students know how the task will be judged? Are they aware of the assessment criteria?
- Does the task contain multiple steps?
- Does the task provide the teacher with the information needed to create the next instructional steps?

FIGURE 4–1. Characteristics of a good classroom assessment task.

the appropriateness, even fairness, of the activity for her students in terms of their past learning experiences. She was concerned with the link between instruction and assessment.

Lucy's point is a good one. It reminds us that language is central to the work of alternative assessment. When the goal is to check whether the student can recall facts or procedures, attention to language becomes less critical. The language used in multiple-choice or fill-in-the-blank items is often abbreviated, intended to elicit a quick, specific response. However, the situation changes when the goal is to procure evidence, in the form of the student's written explanation, that a concept has been understood. Concepts are not discrete facts; they are progressive abstractions of big ideas like numbers, proportionality, and perspective. We form our understanding of and model these ideas through experience and communicate our understanding through written or spoken language. Hence, the activity or problem we use in our classrooms to provoke a demonstration of a student's conceptual understanding is frequently language dependent.

When we want students to demonstrate their mathematical *understanding*, the most appropriate tasks are those that require multiple steps, that require students to decide *what* knowledge to apply *when*. As the study group members noted, the open-ended tasks used as part of their seminars frequently required students to explain or tell how they went about applying their mathematical knowledge. Children's self-assessment of their work also depends on language if that self-assessment is to be shared. These are language-intensive experiences, both for students and for their teachers, and they require:

- Opportunities for students to practice using language, written or spoken, to communicate their mathematical understanding.
- Opportunities to establish a common meaning for requests such as explain your thinking, represent your thinking, and tell me why you solved it that way.
- Opportunities to make clear to your students how their understanding will be evaluated. (Make your assessment criteria and assessment rubrics public; post them in your classroom. Build some of your assessment rubrics in conjunction with your students.)
- Vigilance about checking in with students about their understanding of the language you are using or the language that is emerging in your classroom to describe mathematical ideas.

Communication in mathematics is a meaningful window into students' progressive understanding of key concepts. But as Francine pointed out in our study group conversation, it is important to give students some guideposts along the way. If the communication activities we integrate into our classroom assessment activities are too loosely defined, they will not provide information helpful in teaching mathematics or in documenting student growth in mathematics.

More Ideas for You or Your Study Group

1. Reread the project study group discussion in this session. Then:

    a. Decide whether the characteristics you identified for good assessment problems or investigations would change in any way if the question were, *Can you identify characteristics of good classroom instruction tasks?*
    b. Evaluate the characteristics you feel are the same whether you are thinking of instruction or assessment, as well as those that are different. Why is this so?
    c. Find examples of instructional and assessment activities that contain the kinds of characteristics you have identified.

2. Brainstorm terms or phrases that you associate with classroom assessment activities.

    a. List the terms you come up with.
    b. Define the terms or phrases on the list.
    c. Share your definitions with your colleagues.

    You will undoubtedly find some interesting similarities and differences in these definitions and, as a result, take some important steps toward developing a common understanding.

SESSION 5

Looking at Student Work Across Grade Levels

Background Information

Often, comparisons of student assessments across grade levels have little to do with teaching and learning; rather, such comparisons concentrate on the mastery of mathematical topics at particular grade levels. For example, a second-grade teacher talks about student progress in terms of understanding place value, a kindergarten or first-grade teacher focuses on counting skills. Likewise, in traditional mathematics curriculums, certain mathematical skills—counting, subtraction, division—are covered in particular grades. Such skills-oriented curriculums do little to foster ongoing conversations among teachers across grade levels about how to judge the development of students' mathematical thinking and how to document that development.

Traditional norm-referenced tests also restrict across-grade-level conversations. These tests monitor an array of skills thought to be good predictors of educational success, but they do not provide an understanding of students' mathematical thinking. Standardized test scores are an accountability tool for schools, a means of generating rankings and ratings of student achievement so that schools and school districts can be compared with one another. Whether such student ratings and rankings relate to everyday instructional decisions, however, is debatable.

Walt Haney (1991) comments on the kind of student information teachers value:

> Teachers want assessments that provide them with diagnostic information to help identify strengths and weaknesses in particular students and among groups of students in individual classes and in particular subjects.

> Also, given that teachers teach every day, they want rapid feedback, if not within minutes, then at least within a day or two. This desideratum is generally inconsistent with external testing programs (i.e., testing programs created outside individual schools) in which tests are sent away to be scored by machines or by independent raters. (p. 142)

Jan Mokros, Susan Jo Russell, and Karen Economopoulos (1995) offer a similar perspective:

> Traditional math curricula make it relatively easy to develop quizzes, checklists, and assignments to assess students' learning. Generating test problems by substituting different numbers into problems presented in the textbook, counting how many problems each student gets correct, and determining scores or grades based on this number—traditional assessment often seems that simple. The reform movement in mathematics asks much more of assessment: It asks that teachers truly examine students' mathematical work, question students about their thinking, and observe their strategies for solving challenging, multifaceted problems. This deeper probing of the progress in students' thinking lies at the heart of a constructivist approach. Assessment and teaching depend on the same critical ingredient: a solid understanding of students' mathematical thinking. (p. 84)

Haney and Mokros et al. are advocating alternative assessment practices that result in:

1. Documentation of the progressive development of students' mathematical thinking.
2. An understanding of students' thinking, based on that documentation, that helps shape ongoing instructional decisions, not just at one grade for one teacher, but across grade levels as well.

Manipulating numbers, using problem-solving strategies, judging whether a mathematical solution is reasonable, visualizing spatial relationships, and organizing data are some of the foundational ideas in elementary mathematics. However, these abilities are very different from those usually targeted in specific grades. Mastery of two- or three-digit addition, for example, is a very different type of achievement from an ongoing development of number sense. Understanding how to manipulate numbers is not a rote skill. It is a big, complex idea that needs to be embedded throughout elementary school.

Because number fluency has so many dimensions beyond standard algorithms and memorized facts, the demonstration of this ability will undoubtedly look different in different grades—and even among students at the same grade—and require ongoing conversations among teachers at various grade levels. What does number fluency look like in first grade, in third grade, in fifth grade? What assessment tasks will provide students at each of these grade levels with an opportunity to demonstrate their developing number fluency?

Using alternative assessments allows us to examine the role of professional collaboration in instructional decision making. *Do these students understand data and how they can be organized?* is a very different question from the more traditional *How did these students score on the problem-solving portion of the test?* The first question directs our attention to learning, the second, to issues of scoring and ranking. To piece together ways in which students come to understand increasingly complex data and organize that information in more sophisticated ways, we need to interpret student work, not score it.

Collaborative conversations among teachers at a variety of grade levels are crucial to those interpretations. Without these conversations, there is little hope of unpacking the big curricular outcomes, such as data analysis, identified in the standards and in local state and district curricular frameworks.

Whether a child understands and knows how to use data can certainly be addressed within a single grade level, but it needs to be addressed across grade levels as well. Ultimately the question becomes, *What does a progressive understanding of how to use data look like from kindergarten through sixth grade?* (Or eighth grade or twelfth grade, whatever the pertinent span happens to be.)

Session Particulars

Goals

1. To discuss the work of third, fourth, and fifth graders who responded to a common assessment task, What We Recycle (see Figure 5–1).
2. To practice collaborative conversations about students' progressive understanding of data collection, representation, and interpretation as an example of a curricular outcome in mathematics that is not grade-specific.

Sequence of activities

1. Examine the work samples provided at the end of this book of third, fourth, and fifth graders' responses to the What We Recycle assessment task.
2. Identify specific things you notice about these student samples.

3. Locate evidence in how the students represent their recycling data that indicates different levels of understanding.
4. Identify common responses that may or may not be grade-specific.

Suggestions

If you are working as a study group, break into working groups of two or three people and discuss the question, *What evidence can you find among these What We Recycle work samples of a progressive understanding of how to represent and interpret data?* After twenty or twenty-five minutes, share the results of the small-group discussions as an entire study group.

If you are not part of a study group, address this question yourself. Jot down your observations and refer to them as you read the project study group discussion below.

Project Study Group Discussion

JM: What stands out for you about this work? When you look at these student samples from third, fourth, and fifth graders, what do you notice?

Kate: One thing I noticed right away is that the letters to parents from third graders focused on the steps they took to complete the assignment, the process part. They didn't write about their

What We Recycle

During one week have your students keep track of what is recycled and how much is recycled in their homes. Before they begin this investigation you will want to brainstorm with them questions about how to construct a way to keep track of what is recycled so that everyone's information is organized in similar "units."

For example, you might want to ask how everyone is going to determine what recycling categories to have: paper, plastic, metal, and glass? Or some other kinds of categories? Will students weigh each kind of recycling material? Or will they count? Or will they somehow measure each? Or will they do some kind of combination? What makes the most sense? Why?

When the students have brought their recycling information to school and you have publicly posted the information, have each child build a graph that displays what represents "What We Recycle."

After completing graphs, have your students write letters to their parents describing and interpreting their graphs. Fifth graders should include information about what the graph does not tell about what we recycle.

FIGURE 5–1. The What We Recycle problem.

interpretation of their graphs, their understanding of the information contained in their graphs. Students in the fourth and fifth grades wrote about what their graphs meant.

JM: Kate, could you say a bit more about what you mean by "the process part"?

Kate: There was a lot of, *We did this. Then we did that. And then we did this.* Look at Doug's and Wilson's work [*pp. 151–53*], for example. In her work, LeAnn [*p. 155*] described in great detail the progression of class activities: counting the totals for each group of materials, making a choice about the type of graph, making the graph, and so on. She tells her parents in her opening paragraph, "Let me tell you about it, think of it as a tour in words." So for me she is really at the level of procedure rather than interpretation.

JM: Other observations?

Maxine: Building on what Kate has said, I noticed that the fifth graders were able to draw conclusions; they recognized that larger families have more materials to recycle.

Francine: Was the idea that the amount of recycling was influenced by the size of the family something you had discussed as a class?

Maxine: No, not really. We had all the results posted on the bulletin board, and as they were working in groups together that idea surfaced in certain groups of students. I thought, wow! They really are using their higher-level thinking skills on this.

Sharon: I agree with both Maxine and Kate. The work of the younger ones, the third graders, contained a lot more information on how they went about doing the recycling project, but the fifth graders' writing contained conclusions. What I would add is what I saw in the work of the fourth graders. For example, Lucy went beyond describing the process of the project; she made statements about the data from the recycling project. In the second paragraph of her letter to her mom and dad, Lucy [*p. 157–58*] writes, "I think paper was the most because in most houses people probably read one newspaper a day. Then they recycle it." Later on, when she is describing glass, she writes, "Glass was the least. I think it was the least because most things don't come in glass, they come in plastic." Lars [*p. 159*] makes some conjectures as well.

Gail: I noticed less conjecturing about the results of the recycling project on the part of the fifth graders when compared with the fourth graders. The fifth graders in this sample pretty much reasoned that larger families would recycle more than smaller families. Rita [*p. 161*], who is in fourth grade, attributed the difference between the amounts of recycled glass and plastic to people's behavior. She describes people choosing plastic over glass because glass breaks and small children might possibly hurt themselves. Audrey [*p. 163*], also a fourth grader, thinks the difference in quantity between glass and plastic is the result of more items being packaged in plastic than glass.

Sharon: The fifth graders were asked to write about what information their graphs don't tell them, and there isn't much evidence of their doing that. Carter [*p. 165*] didn't address it at all; neither did Jacqueline or Patrick [*pp. 167–69*].

Gail: I think the kinds of graphs these students have been exposed to is really noticeable. The older grades have definitely been exposed to line graphs, though I'm not sure that's the best vehicle to express the results of the recycling project.

Lucy: A big discussion in my room revolved around how to compare the different categories of recycled items. Some kids wanted to count the papers individually and some did not. And some of the kids said they couldn't do a graph, because they were going to be counting some items individually and some would be in bags. Then someone said we would need to make two graphs, one graph for the things we counted and another graph for the things we measured by bagsful. It was very interesting to see which kids understood that unless there was a standard unit of measurement, it was like comparing apples and oranges. Some kids just didn't see it.

Sharon: How do you get your students to see these key concepts? To say, Now wait a minute, we have five bags and six cans, that won't work?

Jo-Anne: For some kids I really think it is developmental, and you can't force the growth. I handle those times when some children say, Hey, you can't do that! by having a class discussion. Then I find that the next time we run into a similar situation, someone will allude to the first event and say, *Do you remember the time when we were doing recycling?* It's kind of the same thing. They always seem to use that first experience as a sort of marker against which to look at a similar yet new discovery. They keep referring back to the old experience; someone will restate what happened.

JM: Children often grasp the similarity of ideas, but it is difficult to know that, unless they have the chance to communicate their thinking. That is why repeating open-ended tasks in which students explore similar conceptual skills is important. Learning is not always a matter of teaching; teachers need to create conditions that are right for exploring and connecting similar ideas.

Kate: That makes me think of a concept like conservation of numbers. Sometimes no matter what experience the kids have, they don't grasp it. Then, all of a sudden, they get it and we don't know why.

Maxine: The dilemma for the teacher is how much to interject into their students' conversations. My students first came up with the idea of measuring newspaper not by the bag, but by the inch. They decided they would stack up the papers and measure how many inches high the stacks were. I knew they would eventually have to face the problem of constructing the graphs, but I tried hard just to listen to their discussion and not interfere. When *do* you enter the discussion and shape it with your own comments or questions?

Lucy: What if they are going down the wrong path? What then?

JM: I would just let them go; eventually they'll realize that it's not working or that things have gotten too complex. At that point, you can help them find their way back to a less complex place in their discussion or their work.

Gail: That happened in my class. We decided to simplify the counting process when we started talking about counting pieces of paper. The students knew counting individual pieces of paper would take too long. So we ended up putting the paper in bags and counting the bags. When they realized that to create a graph they needed a standard unit of measure, they decided the unit would be a bag. There was a lot of discussion about all of this, but I thought it was pretty important; now they know that when they're collecting data, standardized units are important.

JM: On the basis of these work samples, did students understand that the units of measure needed to be the same in order to make comparisons across the categories of recyclable items: paper, glass, and so on? Did their understanding differ by grade level?

Jo-Anne: I think the fourth and fifth graders saw things in terms of building a unit, the grocery bag. However, the third graders counted individual items. I thought that was a big difference.

Gail: But Javier and Gilian [*pp. 171–73*] are fifth graders, and the unit of measure they selected was the individual item, though how they applied that to newspapers isn't clear. Was an item one section of a newspaper or the paper as a whole? Javier writes about stacks of paper but doesn't define what a stack is.

Vicki: I'm not sure these children, whatever their grade level, understood the relationship between the unit of measure and the comparisons among the recycled items. What if weight had been the unit of measure? I wanted to hear a few more what-ifs like this from the fifth graders.

Gail: My class did talk about weight, not in the way Vicki is suggesting, but they argued that scales wouldn't work because not everyone's scale would be calibrated in the same way. It would only be an approximation, not an accurate measure. *[See Figure 5–2.]* I thought that made good sense.

Maxine: My students focused on the idea of most and least, and came up with reasons why there might be more newspapers than metal containers. It almost became a contest.

Vicki: Students in all three grades were inconsistent about how they labeled their graphs. Gilian noted that one unit on the graph paper equaled ten, but didn't say ten what. And Lucy didn't indicate what the increments of two mean on her graph. Yet Rita and others have labeled the unit of measure clearly.

Sharon: I agree, the inconsistency among the students isn't at one particular grade level, it is across all three levels.

Bonnie: What's bothered me from the beginning about this particular task is, how can the students decide how to work on this problem if they don't know why they're gathering the data? How can they decide in what context to measure or group the different recycling items if they don't know what problem they are trying to solve?

JM: The problem asks students to observe what kinds of items are recycled, construct a unit of measure for these items, and create a graph that shows how the quantities compare. But the problem certainly doesn't have to stop there. For example, someone mentioned earlier that some students realized that the unit of measure—the number of bags versus the weight of the bags—might influence the results. What does anyone else think?

Francine: Bonnie, do you think the problem was too open-ended?

Bonnie: I think it could be. A child's mathematical response might be influenced by his confusion about how to create

WHAT WE RECYCLE

On March 25, the class and I brainstormed ideas for how to handle the project on recycling. The students came up with a list of things that they usually recycle and then categorized them in groups of kinds of materials; paper, metal, glass, plastic. We discussed the possible reasons why the items need to be separated. We determined that items are sent to different places for recycling because the process is different for the different materials. We also tried to think of reasons why this data would be useful. The students said that environmental groups might be concerned about whether households in the area were recycling. The recycling plants themselves might wish to know the potential volume of business.

We discussed the variety of ways we could communicate the amount collected. Students first thought of weighing the amount using pounds. One student said that some materials would weigh less than a pound and we should use a smaller unit of measurement. Using grams in metric measurement might be a possibility. The students thought of holding the items when weighing and then subtracting their own weight or placing the items in a container and subtracting the weight of the container. We took a quick survey about the types of scales the students had at home. Since everyone did not have a scale, we thought of other possible common ways to measure. The height of the pile objects was suggested but many students thought this would present problems. Some students thought that counting the items would work. Then a student said that the sizes of each kind of item varies, such as a greeting card compared to the size of a newspaper. A student figured that we could create a range such as small, medium, large, and extra large. A student remarked that the system was getting too complicated. The idea of common storage containers came up. The kind most readily available were plastic grocery bags. The class agreed that this would be easy to do. Some students questioned whether we would crush the metal or glass. A safety factor came up. We decided to keep it simple and keep everything whole. We agreed that the bags should be labeled and that daily collection is important but daily recording is not necessary. The students are now off with their criteria and letters to their parents explaining the project.

FIGURE 5–2. Gail's summary of her class's discussion of the recycling problem.

appropriate units of comparisons among newspapers, tin cans, and plastic containers.

Kate: I wanted to see what would happen when they couldn't decide on a unit of measure. Some students decided they couldn't compare the items—that's clear from what they wrote. Others drew separate graphs, one for paper and cardboard and one for metal, glass, and plastic.

Bonnie: The problem-solving part of the assignment was interesting: how to measure and what to measure in order to be able to compare. I think my question is, Are there times when problems are so open-ended that our assessments will not be a true reflection of children's abilities?

Vicki: I think teaching style enters into the equation.

JM: The degree of open-endedness for teachers may be a question of comfort and teaching style. If you believe, based on your style of teaching and your knowledge of your students, that an assessment task isn't structured enough or that it's not assessing what you intended, you should feel free to modify it. Looking at work across grade levels gives us the opportunity to discuss this point. If you are going to use across-grade assessment projects to diagnose evolving mathematics abilities in your students as well as their progress in your mathematics curriculum, that will dictate how you structure those projects.

Kate: Math in real life is messy, just as this problem is; our students need practice in making good judgments about what is reasonable in solving a problem.

Lucy: When we talked about good assessment tasks, one thing we said is that a good task needs to pinpoint areas of student weakness, areas that need more practice. I think Bonnie's afraid a problem can be so open-ended that it doesn't diagnose student weakness but instead creates confusion about how to solve it.

Kate: Right, we certainly don't want to go in that direction. On the other hand, we need to go beyond traditional word problems. If we put this problem together in a traditional way, what would it say?

Jo-Anne: I see what you mean. It would be, Joan and Brad took six bags of glass, twenty-one bags of paper, and so on, to the recycling center; which graph depicts these quantities correctly? All the information would be there, and the messiness of working to organize the data would be missing.

JM: I think Bonnie's point is valid and one we need to be thinking about as we explore the purpose and role of classroom assessment. We need to continue to raise these kinds of issues.

Reflections

Student behavior associated with the *progressive development* of mathematical concepts or big ideas is just beginning to be identified. Teachers, teacher educators, and researchers need to create clear, usable descriptions of the progression of characteristics indicating that students are good mathematical problem solvers, have a well-developed sense of numbers, have an accurate concept of spatial relationships, and so on.

Classroom teachers are perhaps the best resource for developing such descriptive statements. As teachers become more accustomed to embedding alternative assessment strategies in their mathematics curriculums, there will be more and more opportunities to observe and gather evidence about student understanding of these important mathematical concepts.

Progress in identifying a range of behavior associated with children's evolving mathematical understanding is unlikely to occur unless teachers have the time and the opportunity to observe, discuss, and reflect on the mathematical work of students at various grade levels. These conversations can be challenging, as we attempt to put aside traditional categorizations of student work as meeting particular grade-level expectations and focus instead on identifying evidence of progressive conceptual understanding: *What does the work of a student with a beginning understanding of the relationship among numbers look like? What does the work of a student with a well-developed understanding of the relationship among numbers look like?*

Even though evolving performance levels will probably keep step with the traditional grade-level progression, they shouldn't be viewed as strictly grade-determined (as in the statement, "Our mathematics curriculum says that we should do place value in second grade," for example). Rather, they should be seen as the path on which students will progress, some moving slowly but methodically forward, some taking a number of steps forward and then retreating a bit before moving forward again.

These ideas, mental images, or written descriptions of what progressive development looks like can be translated into benchmarks or performance indicators to shape teacher judgment of student work as well as future instructional decisions. If we have a clear view of the evolving ways in which students develop a sense of number, then we can begin to build good description-based assessment rubrics to judge students' work in problems focused on demonstrating number sense.

It can be difficult to take the time needed to develop description-based rubrics. Seven or eight years ago when I first began to work with schools on alternative assessment, I was struck by how intent teachers and administrators were on "getting to the stuff on rubrics"—meaning number-based rubrics. Teachers and administrators often decide much too quickly that they need to figure out ways to translate student performance on open-ended tasks into point values and from there into a letter grade. The pressure they feel is understandable. However, this rush to translate powerful portraits of student understanding into number-driven rubrics misses the real strength of alternative assessment as a tool with which to document and diagnose students' developing abilities and on which to base future instructional decisions.

And what about the dilemma Bonnie posed—when is an open-ended task that is being used for assessment too open-ended? Bonnie felt that because the recycling problem didn't specify what units of comparison among the categories should be established, the students' struggle with how to convert cans, paper, cardboard, and plastic into comparable units kept them from demonstrating their ability to build and interpret a graph. However, others in the group were curious, even eager, to see how their students found a way to devise comparable units. (See Figure 5–2.)

The question Bonnie raised does not have a definitive answer. Responses vary, depending on the mathematics curriculum being used; teachers' experience with open-ended problems; students' experience with open-ended problems; the mathematical outcomes that have been identified at given grade levels, some of which may not be aligned with the process of open-ended tasks; the culture of a school or a classroom; and any number of similar factors.

Regardless of the circumstances, it is important that these sorts of professional discussions take place. Through them we get a sense of one another's educational philosophy, teaching style, and opinions about mathematical outcomes for students. We stress the importance of communication between students and teachers in the classroom, and we should also stress the importance of communication among teachers outside the classroom.

More Ideas for You or Your Study Group

1. Think about the issue raised by Bonnie: When is an open-ended task too open-ended? Evaluate differing philosophies and identify guidelines for open-ended tasks or investigations in mathematics.
2. Using these ideas, design an open-ended task that can be integrated into the work of students at a number of grade levels. Be

careful to keep focused on what mathematical concept(s) you want to assess. (The easiest way to begin is to focus on a single concept.)

3. Evaluate a sampling of student work, much as the project study group did. Gather as much evidence as possible about the range of student understanding for the selected concept. Then write a descriptive sentence or two detailing what student understanding looks like at various developmental levels, beginning through well-developed.
4. If your initial open-ended problem does not yield as much evidence about the range of student understanding as you would like, revisit the problem and consider ways to revise it based on what you have learned by observing students doing the task and by discussing their work.

Endnote

During our last meeting as a study group, I introduced a list of questions that had recurred throughout the year and had consistently influenced our discussions. I believe they capture a number of the dilemmas experienced by these teachers and principals as they applied alternative assessment strategies in mathematics and practiced judging the assessments completed by their students. The questions are, in no particular order:

- How do open-ended problems help or hinder our understanding of our students' ability to solve problems, communicate mathematically, and make sense of numbers?
- How can we make changes in the classroom without making them seem like add-ons?
- How does alternative assessment fit into instruction?
- How can teachers communicate these kinds of assessment results to parents?
- How do we involve students in building assessment rubrics?
- When we use rubrics, will students be motivated to go beyond the rubric? How can we not create artificial boundaries for learning?
- Do rubrics provide just another way to categorize student work?
- How do we get children to be invested in their own mathematical learning?

Below is an excerpt from a conversation that took place during that final meeting:

> **JM:** A big question we spent a great deal of time discussing is, *How does alternative assessment fit into instruction*? How have your ideas about that changed since we began these sessions?

Lucy: One of my goals this year has been to use classroom discussion more as a way of letting me know about my kids' mathematical thinking, because I don't think kids always equate discussion with doing math—they think math is working with numbers, using paper and pencil, following along in a book, or filling in a skill sheet. To have spent time talking with children and heard them one by one make a case for his or her solution is really powerful to me. To get kids as a whole class to really look together at mathematics has been exciting; I've felt an excitement in the room about mathematical ideas. Now I think I know how to get those kinds of discussions going, sustain them, and make notes about my observations, but it took practice. The plus is that I am getting more information about what they know when I find ways to make assessment part of everyday math.

Jo-Anne: I wonder if expanding the definition of mathematics to go beyond symbols, to include open-ended discussion, doesn't invite more participation from children. That's what has been happening in my classroom; they are fascinated by the different ways their math problems can be solved.

Maxine: Our work in integrating assessment and instruction has helped me become more discerning about how the children are doing all the time, rather than after an isolated test. I don't think I saw how little information I was getting about students from traditional tests.

Gail: For me, the focus on assessment forced me to look more carefully at exactly what I am teaching and why I am teaching it. I found myself looking at the curriculum not just from my perspective, but from the students'. I know how strange that must sound: it did to me when I said it! That shift in perspective came about because of this group, because we looked at their work—not just the answers, but at the reasoning and thinking behind those answers.

Vicki: This process reinforced how important sharing among teachers is and how little time we get to do it. A lot of people have different ideas and different ways of handling things, and we just don't get a chance to share those things.

Bonnie: I agree with Vicki. I think everyone here has contributed in some way to my thinking about what I teach and how I teach it. It does make a difference when you examine children's thinking rather than respond to answers. That's been a real eye-opener.

These teachers have touched on two fundamental truths about collegial conversation, in this case conversation about the complexities of teaching and learning mathematics. First, new ideas and the courage to try them out in the classroom frequently stem from these conversations. Second, those conversations replicate learning environments that are consistent with what we know about how children learn best—constructing new understanding by actively communicating ideas. That learning principle is appropriate far beyond K–12 education. Adults also need environments in which to interact with ideas and new information, as well as opportunities to apply both the ideas and the information in meaningful contexts.

Forming a study group to explore the role of expert judgment in assessment was an experiment. Over the years I have experienced firsthand many of the shortcomings of traditional professional development or inservice programs. Too often, a complex topic, such as assessment, is covered in an unrealistic amount of time—a total of two release days (often months apart) or a series of unconnected hour-long after-school presentations by a number of different consultants. Even more important, little if any time is set aside to experiment with the ideas introduced in these sessions and then talk about what worked and what didn't. My experiences reinforced my belief that traditional professional development practices are rarely constructed on good principles of learning and that they don't do very much to help teachers deal with the complexities of their profession.

In creating the study group on which this book is based, I was attempting to do several things:

1. Offer a professional development program based on constructivist principles of learning.
2. Address a key aspect of alternative assessment in mathematics: the complexity of forming supportable judgments of students' mathematical work and integrating those judgments in teaching.
3. Narrow the gap between professional development experiences and teachers' everyday experiences in the classroom.

I believe the study group was indeed a good vehicle for accomplishing these goals.

The purpose of the study group—to develop expert judgment in assessing student work—was well served by the process established to accomplish that purpose—providing opportunities to practice assessments and reflect on them. The focus this small group of practitioners directed to a particular aspect of alternative assessment as it related to the work of their students effectively nudged our conversations into an ever-deeper analysis of student work. As the school year progressed, study group members were less inclined to be satisfied with their initial reactions. As a result of our collaborative conversations, most of the group members progressed from making intuitive assessments of student work to making analytic ones and from making quick judgments to making more reflective ones.

Along the way the participants had many opportunities to see how their judgment of student work was aligning with that of their colleagues. In most instances, they were reasonably in sync; and as the year progressed, everyone became more willing to challenge other judgments or ask for clarification. Everyone also began to listen more carefully to the evidence others used to support their judgments.

The study group model of professional development has a great deal to offer teachers. It is a strategy that connects teachers with their everyday work in the classroom and with their colleagues. It can also connect teachers with the same principles of learning they are trying to integrate into their curriculum—principles that stress the importance of creating learning situations that are meaningful to students and provide them with ongoing opportunities to solve good problems, often in collaboration with their peers.

Having acknowledged success, however, I also see areas for refinement:

1. Meeting times have to be sacrosanct. Given already crowded school schedules, there were times when our once-a-month meetings had to be rescheduled. Shuffling dates did interfere with the flow of our work; the group had to take time rediscovering our directional thread when more than three or four weeks elapsed between meetings.
2. Teachers should be encouraged to observe one another's students completing an assessment task that will be discussed at an upcoming meeting. Additional points of view can only enhance the study group conversations. The teachers in our study group often spent a lot of time describing children's behavior during an assessment in order to provide a context for the written work they were sharing. That context would be significantly enhanced by additional firsthand classroom observations.
3. Group leadership should be shared among all members. If the same person is always the leader, a sense of authority comes to be associated with that person, and this can limit the growth of the other group members. While conceptual leadership should indeed rest with whoever has the requisite expertise, facilitating activities, such as keeping the conversations going and identifying ways in which study-group experiences can be incorporated into everyday teaching practices, can be shared. When these activities are shared, the sense of ownership group members feel increases, peer leadership becomes a more comfortable concept, and the perspectives that emerge are richer.

Predictably, as our last meeting ended in June, we wondered what steps to take next. How could we continue what we had accomplished? Our ongoing conversations had altered the way we viewed the role of teachers, the role of students, and the connections among instruction, curriculum, and assessment in mathematics (and perhaps other subject areas as well). It certainly seemed appropriate to think that these conversations were being temporarily suspended, not ending.

The next generation or tier of work we need to do in alternative assessment is to increase teachers' ability to be good diagnosticians of student work and to help their students be good diagnosticians of their own work. Alternative assessment should not be forced into existing school-based accountability practices. Efforts to turn alternative assessment into another way to package traditional student achievement scores will bump up against real limitations, and for good reason.

Alternative assessment is intended to document evolving student abilities. Its evaluative framework is quite different from that of standardized testing, which seeks to rank or assign children rather than diagnose and document their development. Alternative assessment seeks to provide information on students' understanding of multidimensional concepts such as number relationships, spatial relationships, relationships among number operations, and so on; on the ways students apply these multidimensional concepts during problem solving; and on the way students communicate their understanding to others.

The ability to understand mathematical content and to apply that content is what the NCTM standards mean by "knowing and doing" mathematics. Whereas standardized testing allows us to make comparative statements about how students, schools, and school districts score on specific mathematical skills such as computation, number recognition, and number concepts, alternative assessment provides information on an individual student's evolving mathematical understanding, which is unique and may not fit into group and school comparisons. While the purposes of standardized assessment and alternative assessment may be complementary, they are not the same. Exchanging one framework for the other or force-fitting one framework into the other is shortsighted and, in the long term, damaging.

I am not addressing the technical issues of equity, generalizability, validity, or reliability among evaluation practices here. I am simply advocating that alternative assessment have a chance to grow and flourish so that it can do what it's intended to do:

1. Inform teachers, students, and parents about children's progressive understanding of content and the application of that content.
2. Help teachers shape day-to-day instructional and curricular decisions.
3. Give children a more central place in understanding and diagnosing their own learning.

Having teachers collaborate on alternative assessment fits well with the original meaning of *assess*, "to sit down beside" (from the Latin *as* + *sidere*). Opportunities to sit down beside our students to observe and reflect with them on their work and to sit down beside colleagues to discuss that student work are equally important ways to move alternative assessment forward. Productive exchanges about our students' work and their emerging understanding of the complex concepts we want them to know and use as adults are a key ingredient in our efforts to reform mathematics, to maintain a central role for assessment in that process, and to improve teacher professional development practices.

References

Alverno College Faculty. 1994. *Student Assessment-as-Learning at Alverno College*. Milwaukee, WI: Alverno Productions.

Ann Arbor Public Schools. 1993. *Alternative Assessment: Evaluating Student Performance in Elementary Mathematics*. Palo Alto, CA: Dale Seymour.

Badger, Elizabeth. 1992. "More Than Testing." *Arithmetic Teacher* 39 (May): 7–11.

California Assessment Program. 1989. *A Question of Thinking: A First Look at Students' Performance on Open-Ended Questions in Mathematics*. Sacramento, CA: California State Department of Education.

Campbell, Patricia F., and Martin L. Johnson. 1995. "How Primary Students Think and Learn." In *Prospects in School Mathematics*, edited by Iris Carl. Reston, VA: National Council of Teachers of Mathematics.

Clark, David, and Linda Wilson. 1994. "Valuing What We See." *Arithmetic Teacher* 87 (October): 542–45.

Diez, Mary, and C. Jean Moon. 1992. "What Do We Want Students to Know? . . . and Other Important Questions." *Educational Leadership* 49 (8): 38–41.

Haney, Walt. 1991. "We Must Take Care: Fitting Assessments to Functions." In *Expanding Student Assessment*, edited by Vito Perone. Alexandria, VA: Association for Supervision and Curriculum Development.

Herman, J., P. S. Aschbacher, and L. Winters. 1992. *A Practical Guide to Alternative Assessment*. Alexandria, VA: Association for Supervision and Curriculum Development.

Mokros, Jan, Susan Jo Russell, and Karen Economopoulos. 1995. *Beyond Arithmetic: Changing Mathematics in the Elementary Classroom*. Palo Alto, CA: Dale Seymour.

Moon, Jean, and Linda Schulman. 1995. *Finding the Connections: Linking Assessment, Instruction, and Curriculum*. Portsmouth, NH: Heinemann.

National Council of Teachers of Mathematics. 1989. *Curriculum and Evaluation Standards for School Mathematics*. Reston, VA: NCTM.

———. 1995. *Assessment Standards for School Mathematics*. Reston, VA: NCTM.

Ohanian, Susan. 1992. *Garbage Pizza, Patchwork Quilts, and Math Magic*. New York: W. H. Freeman.

Parker, Ruth. 1993. *Mathematical Power: Lessons from a Classroom*. Portsmouth, NH: Heinemann.

Regional Educational Laboratory Program on Science and Mathematics Alternative Assessment. 1994. *A Toolkit for Professional Developers: Alternative Assessment*. Portland, OR: Northwest Regional Educational Laboratory.

Schon, Donald. 1989. "Professional Knowledge and Reflective Practice." In *Schooling for Tomorrow*, edited by T. Sergiovanni and J. Moore. Boston: Allyn and Bacon.

Sergiovanni, T., and J. Moore, eds. 1989. *Schooling for Tomorrow*. Boston: Allyn and Bacon.

Wiggins, Grant P. 1993. *Assessing Student Performance: Exploring the Purpose and Limits of Testing*. San Francisco: Jossey-Bass.

Zessoules, Rieneke, and Howard Gardner. 1991. "Authentic Assessment: Beyond the Buzzword and into the Classroom." In *Expanding Student Assessment*, edited by Vito Perone. Alexandria, VA: Association for Supervision and Curriculum Development.

Session One

A group of 8 people are all going camping for 3 days and need to carry their own water. They read in a guide book that 12.5 liters are needed for a party of 5 people for 1 day. Based on the guide book, what is the minimum amount of water the 8 people should carry all together?

Explain your answer.

If, 12.5 liters are needed for 5 people for 1 day well then 45 liters will be needed for three days for 8 people because for 1 person it was 1.5 which all together equaled 12.5 for 1day, for 8 people to go on a 1 day trip you'd have ~~to~~ bring 15 liters.

Sample 1 student work. Activity by Dr. Judith A. Arter. From A *Toolkit for Professional Developers: Alternative Assessment.* Copyright © 1994. Reprinted by permission of the publisher, Northwest Regional Educational Laboratory.

A group of 8 people are all going camping for 3 days and need to carry their own water. They read in a guide book that 12.5 liters are needed for a party of 5 people for 1 day. Based on the guide book, what is the minimum amount of water the 8 people should carry all together?

Explain your answer.

W W W W W W W W

13.6 136 136 136 136 136 136 13.6

```
  12.5
     3
 +   8
 -----
  13.2
```

liters of water
13.6 l for 8 people to bring camping for 3 days

Sample 2 student work. *Activity by* Dr. Judith A. Arter. *From* A Toolkit for Professional Developers: Alternative Assessment. *Copyright* © 1994. *Reprinted by permission of the publisher,* Northwest Regional Educational Laboratory.

A group of 8 people are all going camping for 3 days and need to carry their own water. They read in a guide book that 12.5 liters are needed for a party of 5 people for 1 day. Based on the guide book, what is the minimum amount of water the 8 people should carry all together?

Explain your answer.

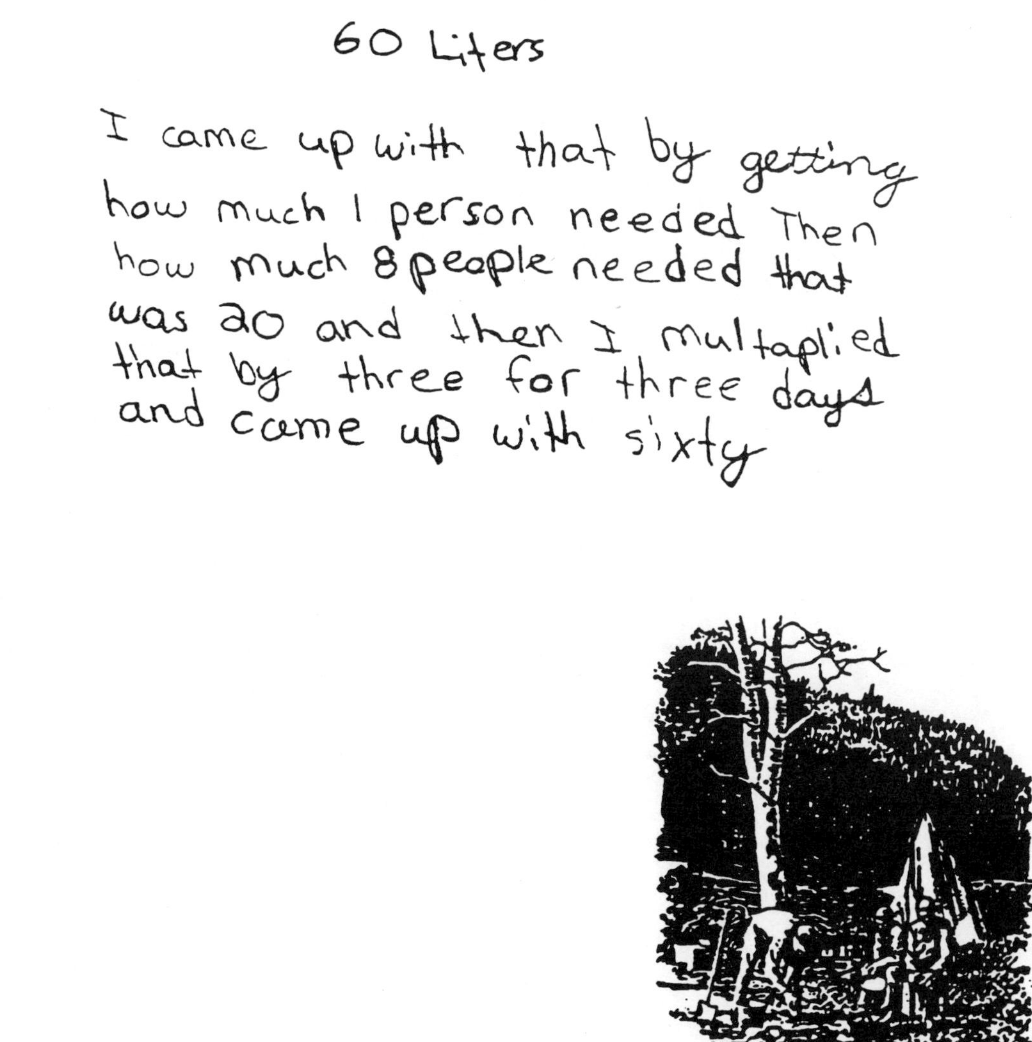

Sample 3 student work. *Activity by* Dr. *Judith* A. *Arter. From* A Toolkit for Professional Developers: Alternative Assessment. *Copyright* © 1994. *Reprinted by permission of the publisher,* *Northwest Regional Educational Laboratory.*

A group of 8 people are all going camping for 3 days and need to carry their own water. They read in a guide book that 12.5 liters are needed for a party of 5 people for 1 day. Based on the guide book, what is the minimum amount of water the 8 people should carry all together?

Explain your answer.

Sample 4 student work. *Activity by* Dr. Judith A. Arter. *From* A Toolkit for Professional Developers: Alternative Assessment. *Copyright* © 1994. *Reprinted by permission of the publisher,* Northwest Regional Educational Laboratory.

A group of 8 people are all going camping for 3 days and need to carry their own water. They read in a guide book that 12.5 liters are needed for a party of 5 people for 1 day. Based on the guide book, what is the minimum amount of water the 8 people should carry all together?

Explain your answer.

5 people) 12.5 liters = 2.5 liters/person

2.5 l/person × 8 people = 20 liters/day

20 liters/day × 3 days = 60 liters in all

I divided 12.5 liters ÷ 5 people = 2.5 liters/person. I did that so that I could take 2.5 liters × 8 peop = 20 liters/day. Now I need to multiply 20 liters/day × 3 days = 60 liters to last the whole camping trip. 60 Liters in all.

Sample 5 student work. *Activity by Dr. Judith A. Arter. From* A Toolkit for Professional Developers: Alternative Assessment. *Copyright* © 1994. *Reprinted by permission of the publisher, Northwest Regional Educational Laboratory.*

A group of 8 people are all going camping for 3 days and need to carry their own water. They read in a guide book that 12.5 liters are needed for a party of 5 people for 1 day. Based on the guide book, what is the minimum amount of water the 8 people should carry all together?

Explain your answer.

Sample 6 student work. *Activity by* Dr. *Judith* A. *Arter. From* A Toolkit for Professional Developers: Alternative Assessment. *Copyright* © 1994. *Reprinted by permission of the publisher, Northwest Regional Educational Laboratory.*

A group of 8 people are all going camping for 3 days and need to carry their own water. They read in a guide book that 12.5 liters are needed for a party of 5 people for 1 day. Based on the guide book, what is the minimum amount of water the 8 people should carry all together?

Explain your answer.

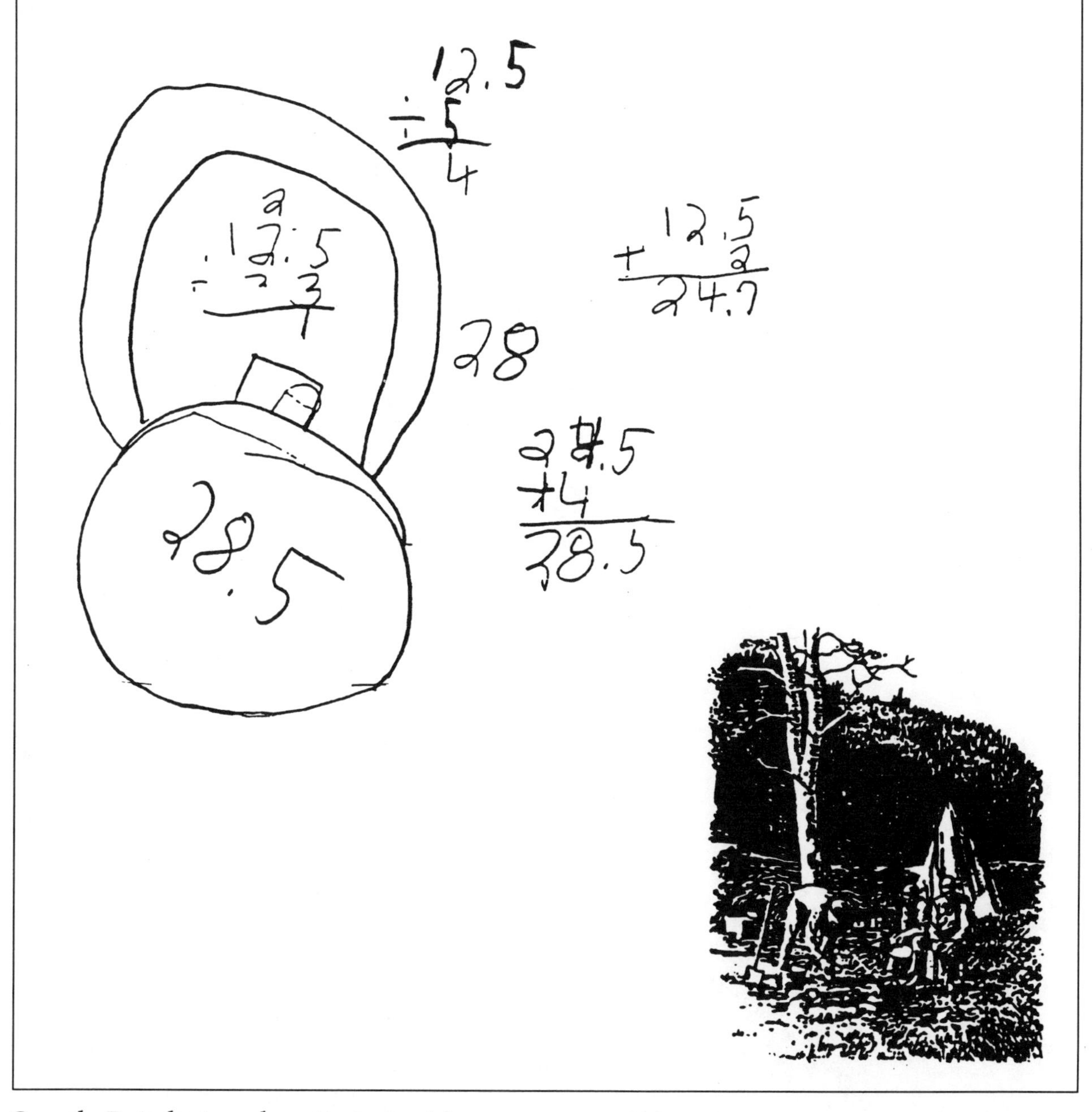

Sample 7 student work. *Activity by* Dr. *Judith* A. *Arter. From* A Toolkit for Professional Developers: Alternative Assessment. *Copyright* © 1994. *Reprinted by permission of the publisher,* Northwest *Regional* Educational *Laboratory.*

A group of 8 people are all going camping for 3 days and need to carry their own water. They read in a guide book that 12.5 liters are needed for a party of 5 people for 1 day. Based on the guide book, what is the minimum amount of water the 8 people should carry all together?

Explain your answer.

Sample 8 student work. *Activity by* Dr. *Judith* A. *Arter. From* A Toolkit for Professional Developers: Alternative Assessment. *Copyright* © 1994. *Reprinted by permission of the publisher,* *Northwest Regional Educational Laboratory.*

A group of 8 people are all going camping for 3 days and need to carry their own water. They read in a guide book that 12.5 liters are needed for a party of 5 people for 1 day. Based on the guide book, what is the minimum amount of water the 8 people should carry all together?

Explain your answer.

2.5 lieters of water apiece
5 people) 12.5 lieters of water for a group of 5 people.

2.5 lieters each of water
x 8 people
20 lieters of water needs to be token at 2.5 lieters each

Sample 9 student work. *Activity by* Dr. *Judith* A. *Arter. From* A Toolkit for Professional Developers: Alternative Assessment. *Copyright* © 1994. *Reprinted by permission of the publisher,* Northwest Regional Educational Laboratory.

A group of 8 people are all going camping for 3 days and need to carry their own water. They read in a guide book that 12.5 liters are needed for a party of 5 people for 1 day. Based on the guide book, what is the minimum amount of water the 8 people should carry all together?

Explain your answer.

12.5
× 2 more people
25
× 2 more days
50

info.
much 12.5
8 people 3 days

First e gathered some important info. 5 peopl for one day was 12.5 e add two people by timing to equel 25. then e times it by two again for the day e new that one day was all ready couned 50 liters

Sample 10 student work. *Activity by* Dr. *Judith* A. *Arter. From* A Toolkit for Professional Developers: Alternative Assessment. *Copyright* © 1994. *Reprinted by permission of the publisher,* Northwest *Regional Educational Laboratory.*

A group of 8 people are all going camping for 3 days and need to carry their own water. They read in a guide book that 12.5 liters are needed for a party of 5 people for 1 day. Based on the guide book, what is the minimum amount of water the 8 people should carry all together?

Explain your answer.

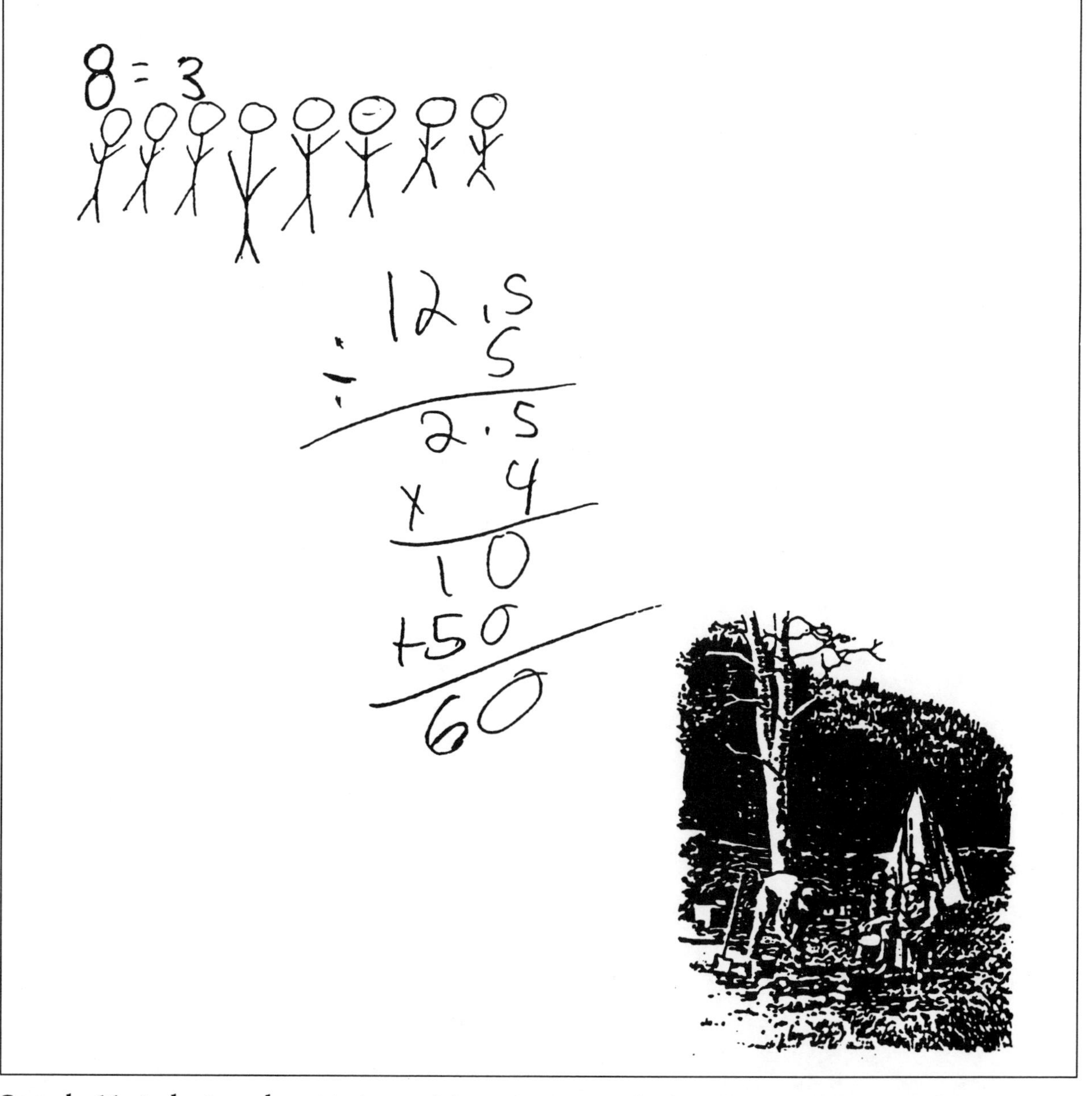

Sample 11 student work. *Activity by* Dr. Judith A. Arter. *From* A Toolkit for Professional Developers: Alternative Assessment. *Copyright © 1994. Reprinted by permission of the publisher,* Northwest Regional Educational Laboratory.

A group of 8 people are all going camping for 3 days and need to carry their own water. They read in a guide book that 12.5 liters are needed for a party of 5 people for 1 day. Based on the guide book, what is the minimum amount of water the 8 people should carry all together?

Explain your answer.

They can bring enough water for 5 people for 6 days which 75 liters which should be enough

Sample 12 student work. *Activity by* Dr. Judith A. Arter. *From* A Toolkit for Professional Developers: Alternative Assessment. *Copyright*

Session Two

In the year of 1994-95 the Fairfield school district made a grapt of their school this year. It shows the months that they recycled and how much they recycled and when they recycled the most and the least.

The Fairfield school district

September
October
November
December
January
Febuary
March
April
May

September October November December January Febuary March April May June

Cathy's work.

When we started school in September, the program started out great we raised 5,000 sheets of paper to recycle. In October the program got better we raised 5,200 sheets of paper. During November it stayed pretty much the same. In December the amont of paper dropped to 2,500 sheets. January was the best worst the paper mill went on strike so we didn't get are paper untill Febuary. March wasn't any better April was the best we raised 7,500 sheets of papper. May and June both stayed at 5000. The averige of the program was norally 5,000 sheets.

8000
7000
6000
5000
4000
3000
2000
1000
0
September October November December January Febuary March April May June

Kate's work.

Developing Judgment: Assessing Children's Work in Mathematics by Jean Moon, © 1997. Portsmouth, NH: Heinemann.

Fairfield Recycling Program

This graph shows that the paper recycled did not have any pattern. It just went up, then down, and down and up. The the most paper recycled was in May and there was about 700 piles of paper in that month. The lowest we had was in Febuary and there was about 150 piles of paper recyled. Ths was one of the programs best schad year!

800
700
600
500
400
300
200
100
Number of piles of paper
Sep. Nov Dec Jan Feb March April May June July
Months

Sam's work.

Paper drive of 1994-95

When school started in September there was about 100 boxes of paper. In October there was a small rise about 5 more boxes then September. I November another rise about 3 more boxes then last month. It went way down to about 48 boxes. January was afull. Only about 30 boxes came in. In Febuary it was way up againto about 130 boxes. March Snow and Ice storms made the number drop to about 70 boxes. April was asome. they got about 150 boxes of paper. In May it droped to about 100 boxes. June was the last moanthe of school. The total boxes of paper in June was 102. That is the report of The Paper drive of 1994-95.

The averige was about 100 papers each moanth. The school learned that sometimes it can be hard and sometimes it can be easy.

Robbie's work.

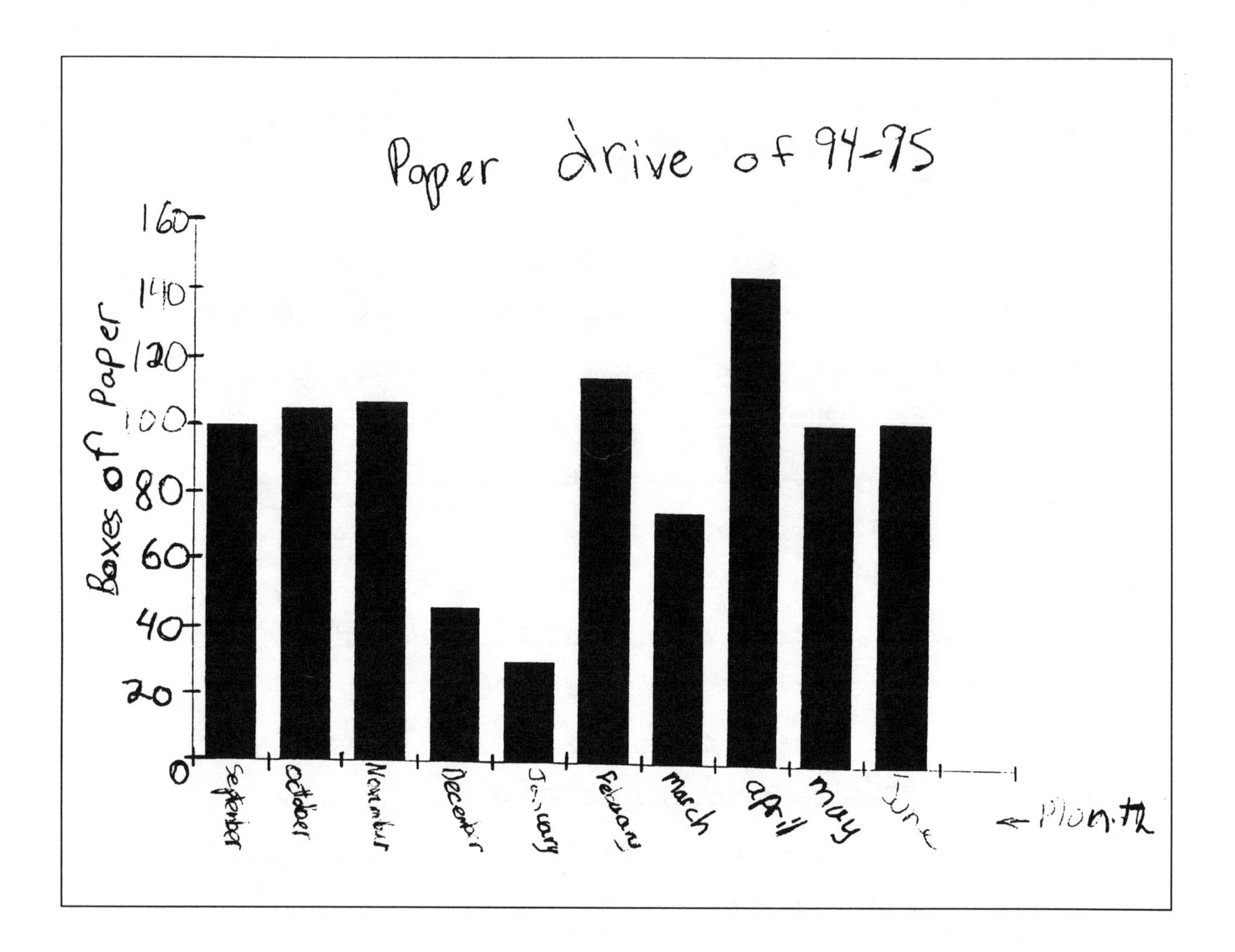

Robbie's work *continued*.

Recycling Paper

First I made a title for the graph.
I then put all the months
except July and August because
you don't go to school.
Then I put, counting by fives,
put the numbers vertical.

Absent Days

40
35
30
25
20
15
10
5

Jan Feb Mar. Apr May June Sept Oct Nov Dec

Darcy's work.

Fairfield

In July the Fairfield District recycled 50 ~~sheets~~ of paper.
In August the Fairfield District recycled 52 sheets of paper
In September the school recycled 53 sheets of paper.
In October Fairfield's ~~recycl~~ recycling program dropped down to 20 sheets of paper.
In November it dropped even lower to 13 sheets of paper.
In December the recycling program rocket lifted to 59 sheets of paper!
In January the school recycled 38 sheets of paper.
In February they went up and recycled 70 sheets of paper that's the most of the year.
In March they recycled 48 sheets of peper.
In April it stayed the same

80
70
60
50
40
30
20
10
0

July August September October November December January February March April

John's work.

Paper Recycling

Between Sep. and Nov. the paper recycling rose, but between Dec. and January it dropped. In Feb. it rose again. Then in Mar. it dropped down again. Apr. rose up to 36, but in May it dropped down to 35. And then it stayed the same in June.

40
35
30
25
20
15
10
5
1

Sep. Oct Nov Dec Jan Feb Mar Apr May Jun

Cassandra's work.

Fairfield Recycling Program

~~The program was successful~~. The recycling was steady. It was in ~~June~~ the spring and summer when the program was most successful. In the winter, it was slow because of the closing of the school for holidays, the paper mill went on strike, and the schools were closed due to snow. In the end, the program was pretty successful. It made a big diffence and it's a start.

70
60
50
40
30
20
10
Sept. Oct. Nov. Feb. Mar April May June

Daniel's work.

12-12-95 Math

The Fairfeild school recycling program did the most recycling out of all the thirty schools that participated. Sue Lau is in the fifth grade at Fairfield school. Sue made a graph of how much their school recycled. She got a little imormation from the recyling person. Sue found out that the most recycling was in April. They recycled 400 pounds in April. The least they recycled was 100 pounds in January. In the graph Sue measured by 50ds. She started with 50 pounds and ended with 400 pounds. When Sue was making the graph she wrote down how much each month recycled. This is what it looked like: January 100 LPS, December 150 LPS, March 250 LPS, September 300 LPS, May 300 LPS, June 300 LPS, October 300 LPS, November 300 LPS, February 350 LPS, April 400 LPS. Sue wanted to do a bar graph. Now Sue's bar graph is hanging in the awards hall at Fairfield

Kendra's work.

school. Sue is now in sixth grade at the Winter, Junior High. Sue Lou started her new school in the, recycling program. Instedd of just recycling paper all the schools in the program recycle everything. The list that Sue made was even recycled. The princapal had to make a list of stuff you can bring in cause she foand too much weird stuff coming in. The list was: Paper, glass bottles, plastic and other unweird things.

450 LBS
400 LBS
350 LPS
300 LPS
250 LPS
200 LPS
150 LPS
100 LPS
50 pounds

Sep. Oct. Nov. Dec. Jan. Feb. Mar. Apr. May June

Kendra's work ***continued.***

Session Three

List some ways that these graphs are alike.
Tell a classmate why you think so.

They are ~~bo~~graphs. They are the same in many ways. They are showing how mach food is wasted in both catagorys.

Andy's work in November.

Tell all you can about these 2 graphs. How are they alike? How are they different?

These graphs are different because in Feb. for My class there is no body but at winslow there is two poeple on Feb. In april for winslow there is no body on Apr. & my class there is tree poeple on Apr. In Aug. there is two people at winslow and two people at my class too. I Mar. for winslow there is three people and for my class there is no body on march In my class therare two peele on Dec. and for winslow there are one person on the Dec. In my class there are 24 poele in winslow there are 20 poele in the whole class. In winslow June there are 1 persun and for my class there are 4 people.

Andy's work in March.

List some ways that these graphs are alike.
Tell a classmate why you think so.

These ghaphs are alike because they show us how much food was wasted, and how much food they did wast, and what kinds of food they did wast. This boy wasted one ice creamcones This girl wasted eight cartons of milk. This boy wasted three hot dogs. This girl wasted three pears.

Samantha's work in November.

Tell all you can about these 2 graphs. How are they alike? How are they different?

Alike... They both have 1 bithday on october. They both have three bithdays on september. They both have two bithdays in july. They both have two bithdays in augest.

Diffrent... One graph has no bithdays on april and the other has no bithdays on March. One graph has two bithdays on december and the other has one bithday on december. One graph is a bar graph and the other graph is not.

Samantha's work in March.

Session Five

Dear Mom & Eric

1. We descused how to count.

2. We brought the tabels home for the 5 days and counted them up at home

3. Kids (thats me) added up the total for each of them.

4. We did more talking!

6. We voted!

(and last but not least the end)

7. We made a Class graph

Love & your buddy

Doug's work.

April 8, 1996

Dear Parents

Mrs. Wool asind a pojet to colecte datae ate home. We talked alote about how to count the things we recycling. Mrs Wool gave us talles to take home for 5 days. Kids added up a total of eache product. the totals were metal with 128, paper with 167 plastic with 157 and glass with 119. Then we talked more. We tookk a vote on which kind of graph we wanted to use a pictograph or a bare graph. We made a class graph.

Wilson's work.

Dear Parents,

Our five day recycling project is Complete. Let me tell you about it, think of it as a tour in words.

Every student was in charge of his/her recycling. The things we recycled were, Metal, Paper, Plastic, and glass.

On Tuesday we had 4 groups of people count up the total of of the 4 choices.

Next came making a graph to see all the informaiton.

The question was should we do a bargraph or a pictograph?

We voted and piclograph won. But after Mrs. Wool made on them, we decided to that the bargraph was esier to read, because we counted by 10's and if an sary plastic was 154 you could tell that it was 154. And that's our project in words.

Metal Paper Plastic Glass
178 167 157 119

LeAnn's work.

Dear Mom and Dad,

I'm writing home to tell you about my recycling project. Rember how I had all those bags and I was measuring things. That's what i'm talking about.

We recycled the most paper. (31 bags to be egsacked). I think paper was the most because in most houses people probably read one newspaper a day. Then they recycle it. One newspaper takes up half a bag! Plastic was the next most. I think ~~plastic~~ was 27 bags because soda, milk, and orange juice. Those all come in big plastic bottels.

Metal was next. I think there was quite a bit of metal because soda cans are metal. People drink alot of soda cans.

Glass was the least. I think it was the least because most things

don't come in glass. They only come in plastic. We only had 15 bags of glass. I didn't recycle any glass. Believe it or not I recycled everything in the same order. Paper the most, then plastic, then metal, than glass. Thanks for your cooperation.

Lucy's work.

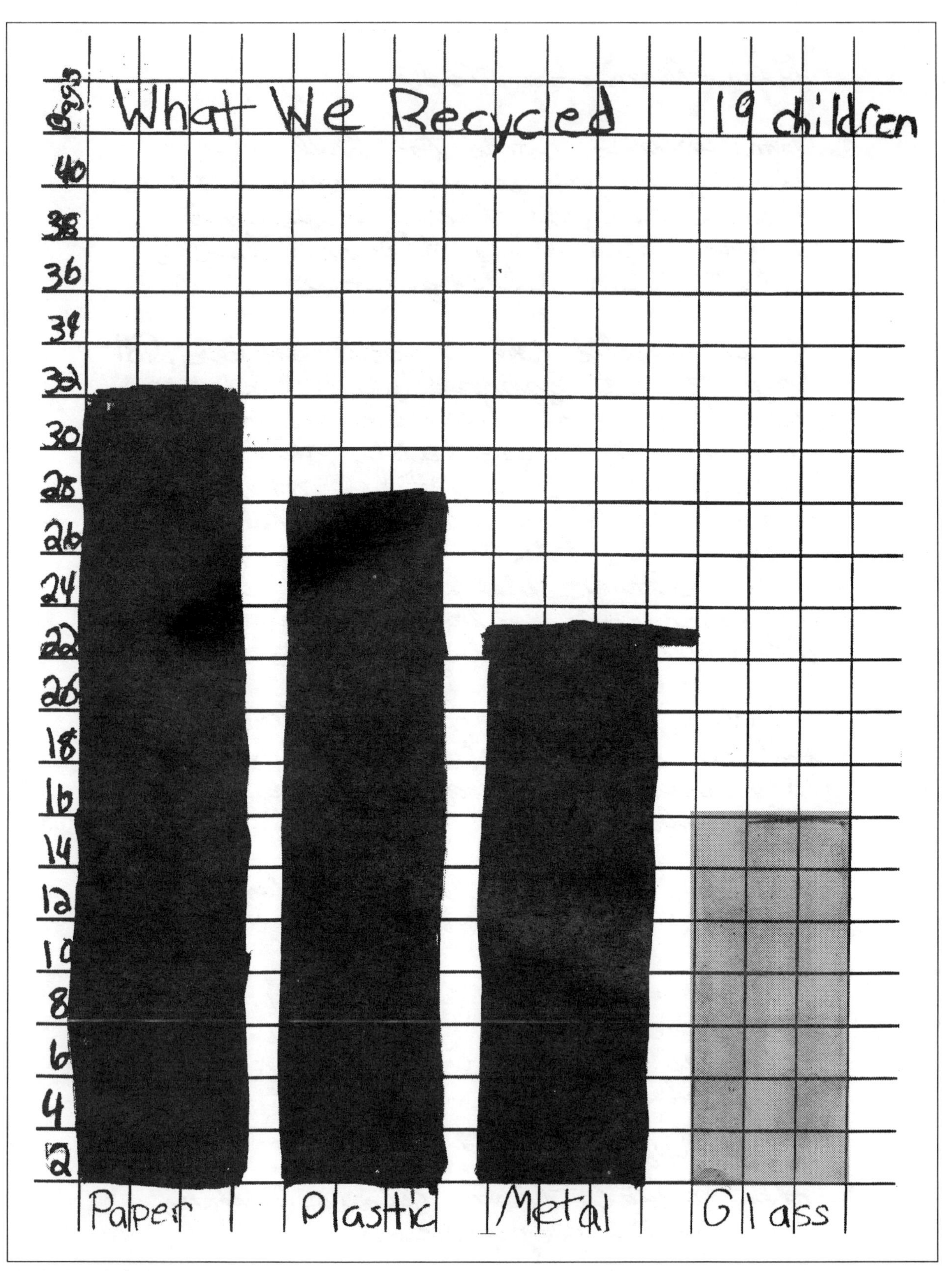

Lucy's work *continued*.

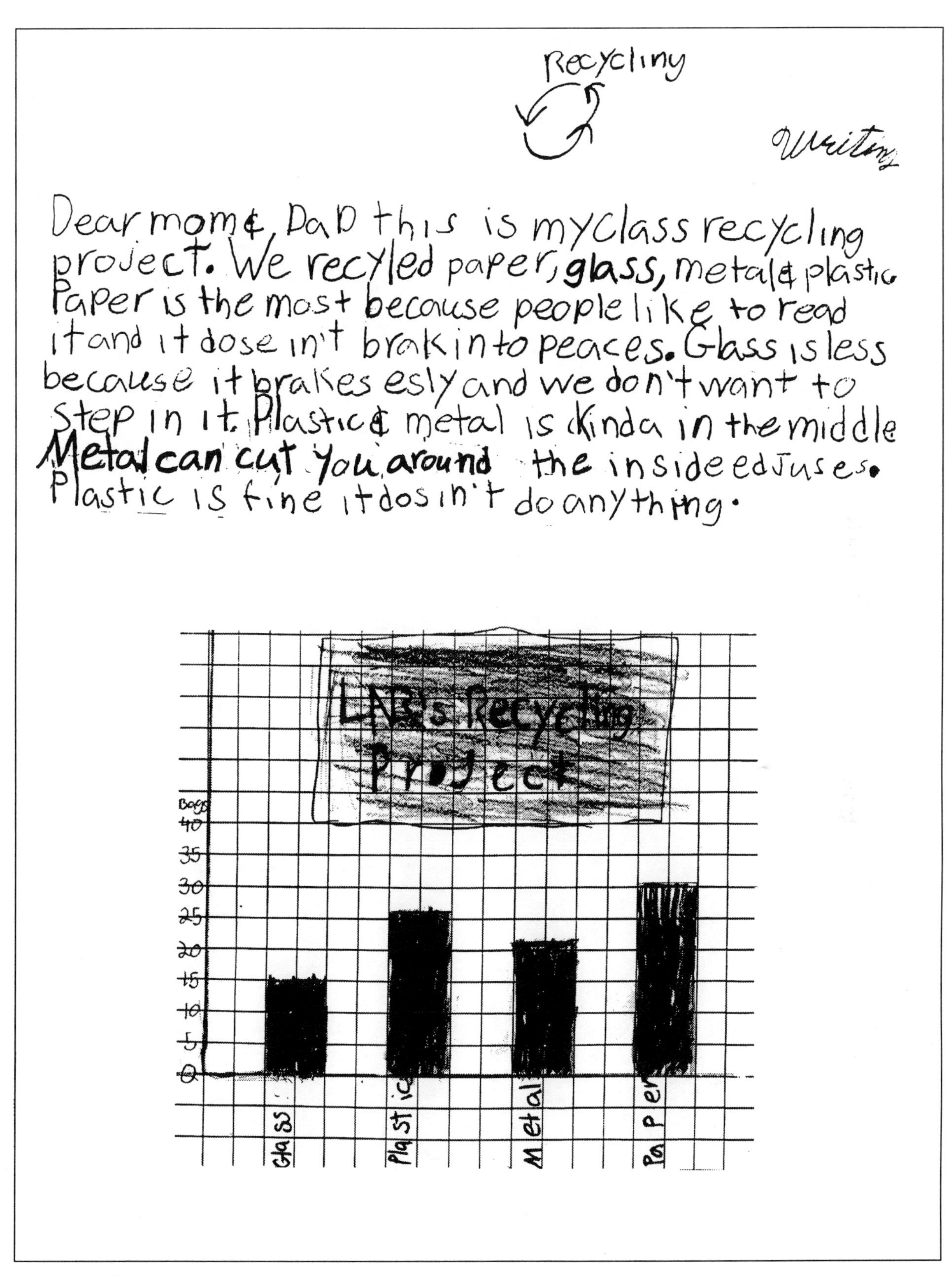

Recycling
Writing
Dear mom & Dad this is my class recycling
project. We recyled paper, glass, metal & plastic
Paper is the most because people like to read
it and it dose in't brak into peaces. Glass is less
because it brakes esly and we don't want to
step in it. Plastic & metal is kinda in the middle
Metal can cut you around the inside edjuses.
Plastic is fine it dosin't do anything.
LArs's Recycling Project
Bags
40
35
30
25
20
15
10
5
0
Glass
Plastic
Metal
Paper

Lars's work.

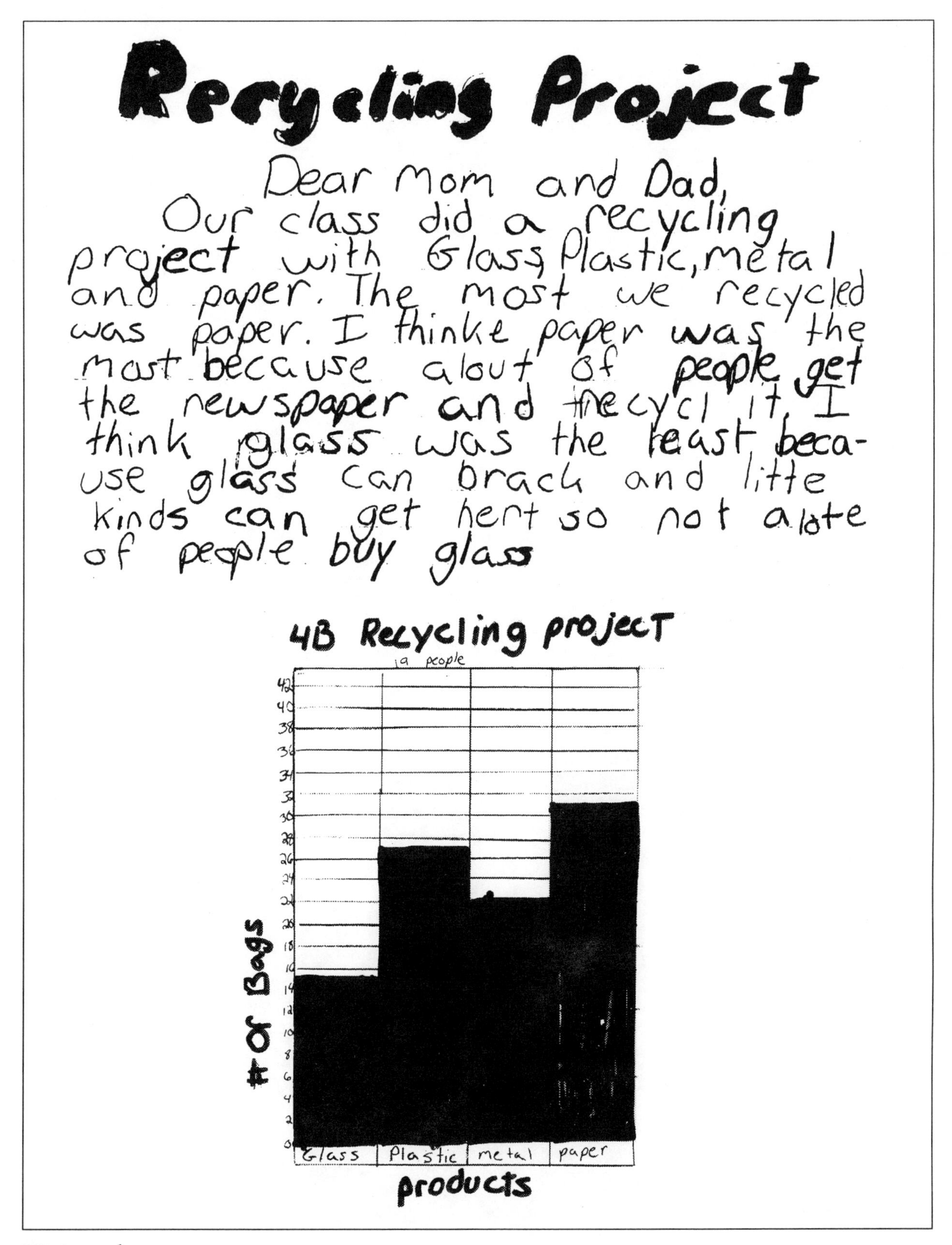
Recycling Project

Dear Mom and Dad,
Our class did a recycling project with Glass, Plastic, metal and paper. The most we recycled was paper. I thinke paper was the most because alout of people get the newspaper and recycl it. I think glass was the least because glass can brack and litte kinds can get hert so not alote of people buy glass

Rita's work.

Dear Mom and Dad,

The results of our graph were as follows. Paper :31 bags Plastic :27 bags Metal :22 bags Glass :15 bags.

I think paper is the most because most people recive the newspaper every day. I think plastic was the second most because everthing seems to be made of plastic now a day. Such as packaging Materiels, and Soda bottles. I think glass is the least because everthing is made of plastic instead.

Recycling Chart

Paper
Plastic
Medle
Glass

2 4 6 8 10 12 14 16 18 22 24 26 28 30 32 34

Audrey's work.

Dear Mom and Dad, 4/8/96

I learned that people in our Class recycle paper the most, and glass the least. Kids with bigger families tend to recycle more, and kids with smaller families recycle less.

	CANS	Glass	PAPER	Plastic
	266	148	253	201

■ = 50 pieces of recycling

Carter's work.

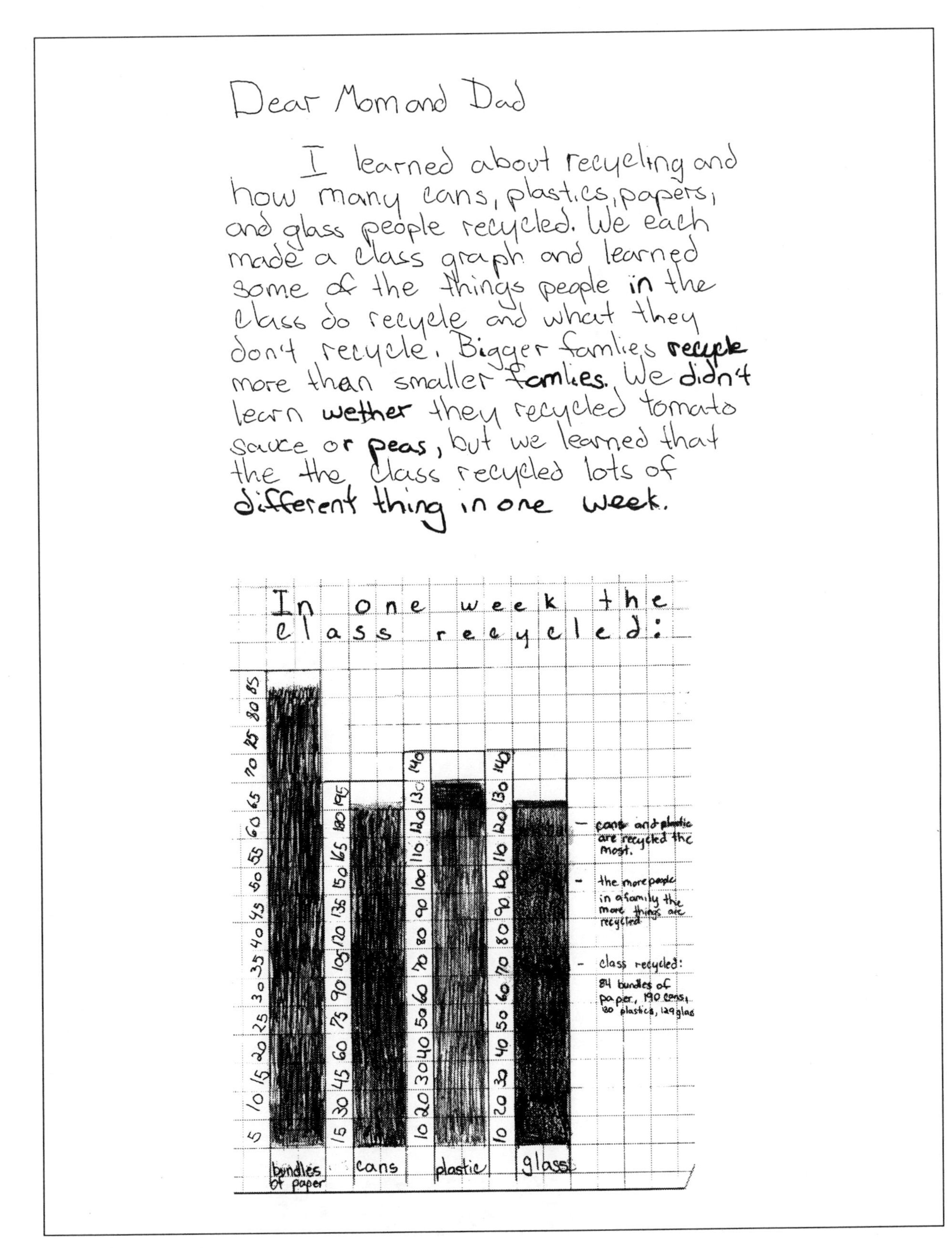

Dear Mom and Dad

I learned about recycling and how many cans, plastics, papers, and glass people recycled. We each made a class graph and learned some of the things people in the class do recycle and what they don't recycle. Bigger famlies recyple more than smaller famlies. We didn't learn wether they recycled tomato sauce or peas, but we learned that the the class recycled lots of different thing in one week.

Jacqueline's work.

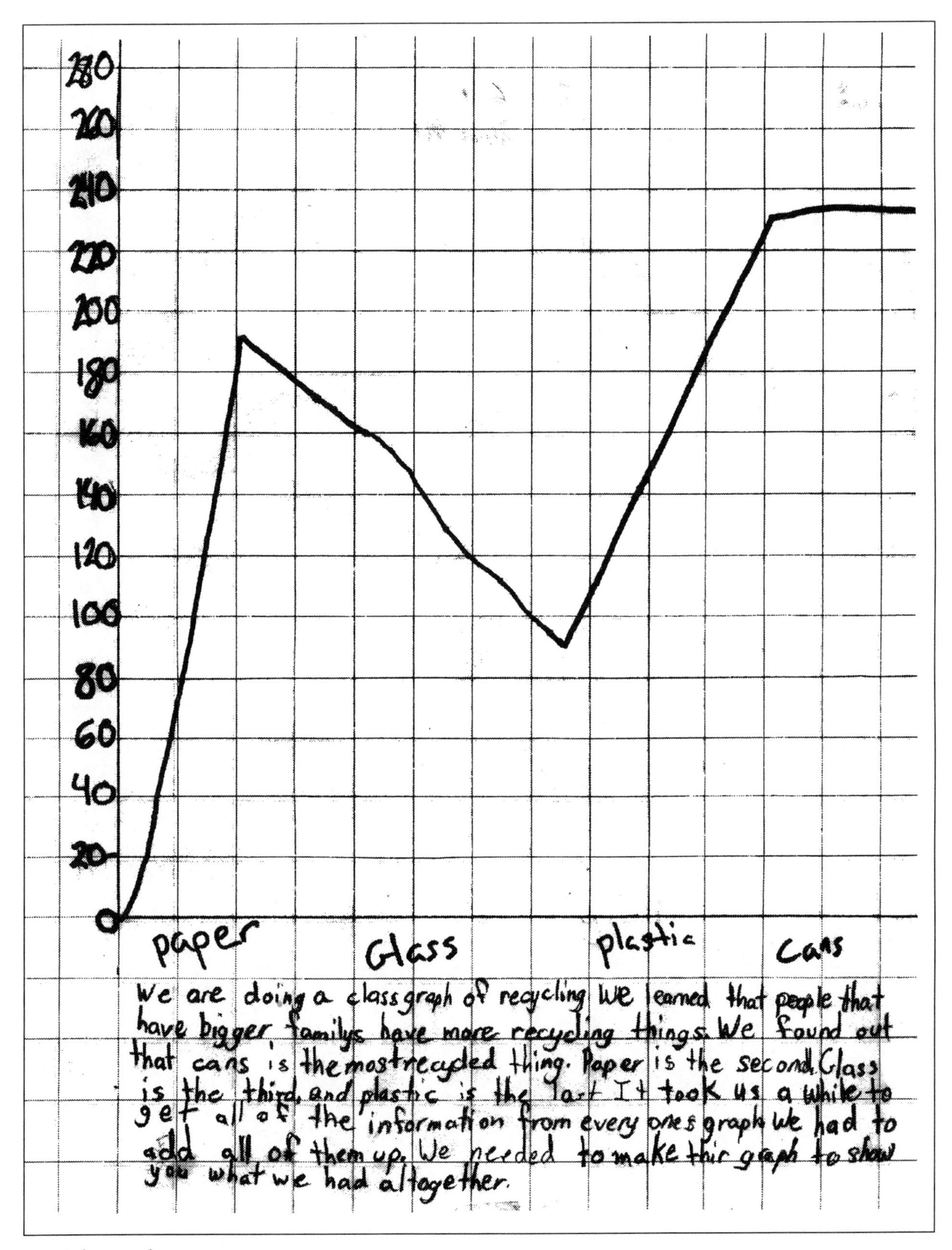

Patrick's work.

Dear Mom and Dad,

Over the last week we graphed what we recycled. We recycled paper, plastic, glass, and tin. We recycled about 275 stacks of paper, 225 plastic containers, 100 glass jars, and about 310 pieces of tin. We ~~rcv~~ recycled the most of tin and the least of glass. We recycled a lot of things over the last week.

300
250
200
150
100
50
0

Paper Plastic glass tin

Javier's work.

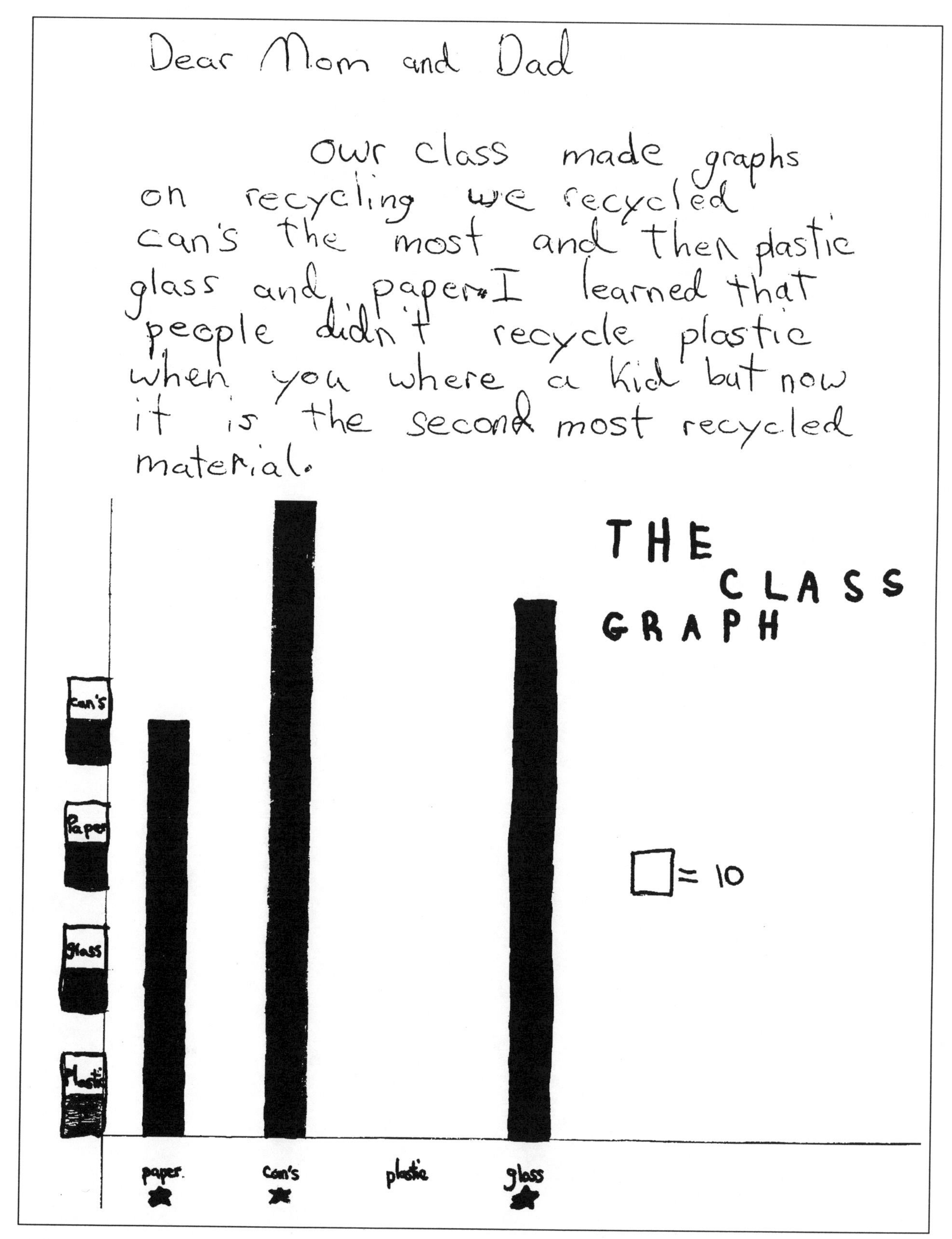

Gilian's work.

Developing Judgment: Assessing Children's Work in Mathematics by Jean Moon, © 1997. Portsmouth, NH: Heinemann.